Lecture Notes in Engineering

Edited by C. A. Brebbia and S. A. Orszag

68

T. G. B. DeFigueiredo

A New Boundary Element Formulation in Engineering

Springer-Verlag
Berlin Heidelberg New York
London Paris Tokyo
Hong Kong Barcelona Budapest

Series Editors
C. A. Brebbia · S. A. Orszag

Consulting Editors
J. Argyris · K.-J. Bathe · A. S. Cakmak · J. Connor · R. McCrory
C. S. Desai · K.-P. Holz · F. A. Leckie · G. Pinder · A. R. S. Pont
J. H. Seinfeld · P. Silvester · P. Spanos · W. Wunderlich · S. Yip

Author
Dr. Tania G. B. DeFigueiredo
Departamento de Engenharia Civil
Universidade de Brasilia
Campus Universitario – Asa Norte
709 10 Brasiliaia-DF
BRAZIL

ISBN 3-540-54030-X Springer-Verlag Berlin Heidelberg New York
ISBN 0-387-54030-X Springer-Verlag New York Berlin Heidelberg

Typesetting: Camera ready by author
Printing: Mercedes-Druck, Berlin
Binding: B. Helm, Berlin
61/3020-543210   Printed on acid-free paper.

To my mother

Mrs. Ivani Ribeiro Alla

ACKNOWLEDGEMENT

I would like to express my gratitude to everyone who contributed, in different ways, to the completion of this work. Inevitably some names will be missing here.

My special thanks go to Dr. C. A. Brebbia for his valuable and friendly supervision and also for his overall support.

I also would like to greatly acknowledge the help from Dr. J. C. F. Telles in the beginning of this research; the great support from Dr. M. H. Aliabadi, with suggestions and discussions throughout this work and also the reading of part of the manuscripts and the encouragement and discussions with Dr. L. C. Wrobel.

I am indebted to Dr. T. G. Phemister, for his kind and spontaneous help with the singular integrations, without which the completion of this work would have to be delayed. I also wish to thank Dr. G. T. Symm for his help as a mathematician.

I am grateful to all the staff of the Wessex Institute of Technology for their support. To Dr. W. Blain for his general assistance, related to my PhD studies. To Miss R. Barrett for her efficient help with accommodation for me and my family. To Dr. S. M. Niku for technical discussions. To Mrs. A. Lampard for her continuous kindness and to Mr. R. Davies for his assistance with the computer equipment.

The useful discussions and suggestions from the Wessex Institute of Technology students, especially J. P. Azevedo and K. Hayami, are greatly appreciated.

The support of the University of Brasília and in particular of the Civil Engineering Department and his former head, Mr. M. A. A. de Souza and the financial assistance from CNPq – Brazilian National Council for Scientific and Technological Development – are gratefully acknowledged.

Finally my gratitude goes to my family: to my husband, Dr. D. De-Figueiredo, for his encouraging support and help, both at home and in my work, and to my children Dimitri, Danielle and Danilo for their mature understanding and respect for my professional career and their overall help.

Contents

Chapter 1

Introduction

1.1 The Hybrid Displacement Boundary Element Model

This work is concerned with the derivation of a numerical model for the solution of boundary-value problems in potential theory and linear elasticity.

It is considered a boundary element model because the final integral equation involves some boundary integrals, whose evaluation requires a boundary discretization. Furthermore, all the unknowns are boundary variables. The model is completely new; it differs from the classical boundary element formulation in the way it is generated and consequently in the final equations. A generalized variational principle is used as a basis for its derivation, whereas the conventional boundary element formulation is based on Green's formula (potential problems) and on Somigliana's identity (elasticity), or alternatively through the weighted residual technique.

The multi-field variational principle which generates the formulation involves three independent variables. For potential problems, these are the potential in the domain and the potential and its normal derivative on the boundary. In the case of elasticity, these variables are displacements in the domain and displacements and tractions on the boundary. For this reason, by analogy with the assumed displacement hybrid finite element model, initially proposed by Tong [1] in 1970, it can be called a hybrid displacement model.

The final system of equations to be solved is similar to that found in a stiffness formulation. The stiffness matrix for this model is symmetric and can be evaluated by only performing integrations along the boundary. The unknowns are boundary potentials (potential problems) or boundary displacements (elasticity problems). The other fundamental variables can be obtained from them.

In general, domain sources (potential problems) or body forces (elasticity) are present in the formulation. In this situation, domain integrals of exactly the same type as those present in the conventional boundary element method appear. Another similarity between the conventional boundary element method and this hybrid model is that the same kind of singular fundamental solutions is used in both approaches. In these two formulations, the use of such fundamental solutions requires the integration of singular functions and, therefore, the application of special integration schemes.

In the following three sections, some historical notes on variational principles, the finite element (including hybrid models) and the boundary element methods will be presented. All these subjects are important when establishing the theoretical foundations for the generation and development of the proposed model. A discussion on boundary element variational formulations will also be presented in section 1.5.

1.2   Historical Development of Variational Principles

In this section, a brief historical account of variational principles in solid mechanics is given. Some of the first and best known variational principles which have appeared in the literature are mentioned. Special reference is given to those which have played a relevant role in the development of numerical techniques, mainly in the case of the finite element method.

Mathematical physics is a branch of science where variational principles has been successfully applied. This is because the behavior of physical systems is often determined by the stationary conditions of a functional. In other words, it is often possible to find a functional whose stationary conditions produce the equations governing the physical phenomenon. This functional sometimes has a physical meaning, such as the case of the time required for a ray of light to travel between two points (Fermat's principle) in optics.

Variational principles have played an important role in the development of many branches of physics. Mechanics is one of its most fruitful areas of application, wherein variational techniques have been extensively investigated and used. The classical problem of particle mechanics, for example, can be expressed either through Newton's law of motion or through its variational form, known as Hamilton's principle. Although both approaches lead to the solution of the problem, the latter often offers advantages when studying complicated physical systems. Problems in dynamics and statics of rigid bodies, general elasticity, the theory of plates and shells, vibrations and many others have been solved using variational principles. Another interesting feature is that they are suitable for determining approximate solutions and, therefore, they provide a basis for the derivation of numerical methods in engineering.

Rayleigh used variational methods for the purpose of obtaining successive approximations to both boundary and eigenvalue problems. This discovery was published in 1870 in the "Philosophical Transactions of the Royal Society", and afterwards, in his "Theory of Sound" [2]. It was, however, only after the work of Ritz that variational principles gained prominence in applied sciences. In Ritz's papers [3, 4], published in 1908 and 1909 the work of Rayleigh was considerably generalized and clarified. He applied the variational method to the determination of natural frequencies and mode shapes of vibrating plates. The method, nowadays known as the Raleigh-Ritz or simply the Ritz method, has been applied in the solution of problems in many different branches of science and engineering.

In 1943, Courant [5] pointed out the connection between the partial differential equation and the variational problem and presented his approximate solution to St. Venant torsion problem. His solution was formulated by the principle of minimum potential energy and it is sometimes considered the first finite element analysis of solid continua.

Hellinger [6], in 1914, presented a variational theorem for finite elasticity problems, with independent displacement and stress fields. The works, which have been mentioned previously, included only one independent variable and hence Hellinger's principle seems to be the first variational principle involving more than one field variable. In 1950, Reissner [7] published a variational statement, for linear elasticity, which involves displacement and stress as independent fields and also includes the boundary conditions. This principle, usually called Hellinger-Reissner or simply Reissner principle, appears to be the first general variational theorem published. It has been extensively used in the development of new finite element formulations, specially multi-field finite element models. An even more general variational statement was independently published in 1955 by Hu [8] and Washizu [9]. This principle, now known as the Hu-Washizu principle, includes independent displacement, strain and stress fields.

Several other single or multi-field variational principles have been proposed since then for linear and nonlinear elasticity, problems of small and large deformations, plasticity, elastodynamics and others. A summary of various generalized variational principles for nonlinear, incremental and linear elasticity is presented in reference [49].

Several general and systematic derivations of variational principles have also been published. The first general way of deriving variational principles with discontinuous fields was presented by Prager [11, 12]. Washizu [13] showed how the principle of virtual work and related variational principles in elasticity and plasticity can be derived. Pian in reference [22] presented the derivation of some extended variational principles for linear solid mechanics. He also described a general functional which involves five different field variables and is the most general functional for finite element formulations. More recently Felippa [41, 42] proposed his parametrized multi-field variational principles for linear elasticity. In his papers, one-parameter families of mixed variational principles are constructed. According to the value of the parameter, several mixed and hybrid variational principles can be derived. The potential energy and the generalized Hellinger-Heissner principles are included among them, as special cases.

Variational methods can also be applied to fields other than structural and solid mechanics, although one of its most successful areas of application has been linear elasticity. Thus problems in steady state and transient heat conduction, fluid mechanics, elasto-dynamics, creep and many others have been solved through the use of variational principles.

1.3 Variational Principles and Finite Element Models

This section contains a short review of the beginning of finite elements, the influence of variational principles in their development, a brief definition of the main finite element models and the connection between these models and variational principles.

The finite element method for numerical analysis of continua began with the work of structural analysts who, guided by their physical understanding of the behaviour of structures, viewed a continuum as an assemblage of discrete structural elements, connected at a finite number of nodes.

At that time, the application of fundamental energy principles of elasticity by Argyris and Kelsey [14] had made possible the development of a general formulation of matrix structural theory. They showed the possibility of choosing displacements, rather than internal forces, as primary unknowns. Consequently the usual stiffness matrix had been brought into the formulation, although it was obtained through the inversion of the flexibility matrix. This development proved to be very important in the formulation of the finite element method. The first direct derivation of a stiffness matrix for a plane triangular element, based on assumed displacements, by M. J. Turner, R. W. Clough, H. C. Martin and L. J. Topp [15], is regarded as the starting point of the finite element concept in its modern form.

It was, however, several years later that studies in convergence of solutions led to the establishment of the finite element method on a variational basis. This fact turned out to be very important for the further development and extension of the method to fields other than structural and solid mechanics.

The first author to present the finite element method on a variational basis was Melosh [71] who, in 1963, presented a finite element formulation based on the principle of minimum potential energy. Subsequent works demonstrated the rich theoretical basis offered by the variational principles for developing new finite element models. In 1964, Jones [17] pointed out the advantages in using Reissner's general variational principle. This fact led to the development of mixed models based on that variational statement. The same year, Pian [18] proposed the first hybrid finite element model based on a modified variational principle, although at that time this fact was not fully understood.

Variational principles can not only be used to generate the formulation, but they can also provide a means to prove the convergence of finite element solutions. In the case of linear solid mechanics, they allow the establishment of upper and lower bounds on the total strain energy ; hence they also serve as an assessment of the accuracy of the solutions.

Scientists working on finite elements generally agree that establishing the method on a variational basis led to advances that would have been impossible otherwise. Pian, in reference [19], states that the wide expansion of finite elements in solid mechanics is obviously due to the emergence of variational principles. Certainly the variational approach made possible a better understanding of the first mixed/hybrid models and also the development of new ones. Martin, in reference [92], comments that the variational approach of finite elements has been responsible for the application of the method to fields other than solid mechanics. In fact, once the physical concept of a finite element was generalized to a mathematical one, the method could be easily extended to problems such as fluid flow, heat transfer and many others. It was no longer necessary to define a physical fluid or temperature element in other to apply the method in these cases.

For these reasons, although finite element formulations need not be based on the variational approach, the use of variational principles in finite elements played an important role in the success and wide development of the method.

According to the variational principle used, and the consequent number and type of independent variables involved in the functional, different single-field or multi-field models are generated. Finite element practice, however, continue to be primarily based on one-field formulations in spite of some shortcomings of conventional single-field element models. This is mainly due to the simplicity of the one-field formulations.

The two basic variational principles in structures and mechanics of solids are the principle of minimum potential energy and the principle of minimum complementary energy. The corresponding one-field models, sometimes referred to as primal finite element models, are the assumed displacement and the assumed stress models. The first is also called the compatible displacement model; it has the displacement as independent field and is the most commonly used. In the second, also called the equilibrium model, the independent variable are the stresses.

Multi-field-models contain more than one field variable in an element. They are generated by using modified variational principles, which in turn are derived by including in the functional the conditions of constraint and the corresponding Lagrange multipliers. These principles are usually stationary principles, not minimum nor maximum principles. The multi-field models are called mixed or hybrid models. The motivation for their early development was the difficulty in constructing a compatible displacement field for problems such as plate bending, which are governed by fourth order differential equations. Later hybrid models also proved to be very accurate when used to solve problems governed by second order equations.

A brief explanation of the main features of the different multi-field models commonly used will be given here. For more complete descriptions see references [21, 22, 23]

Mixed formulations include more than one independent field variable. These variables are defined within the element as well as on its boundary. In structural and solid mechanics problems, the independent variables are stress and kinematic variables, such as displacement or strain. They can be derived either through variational principles or weighted residual approaches. The latter is more general than the former, because it does not require the existence of a variational principle. In solid mechanics, the Hellinger-Heissner principle is the one which is usually used to generate mixed formulations, although the Hu-Washizu principle can also be used.

For hybrid models, some of the field variables are defined within the element and others on its boundary. They are obtained from general variational principles such as the Hu-Washizu principle in solid mechanics, in which relaxing interelement continuity requirements are introduced through Lagrange multipliers. Although the development of element properties starts by approximating more than one field, it is possible to eliminate all but one variable in the final equations, which are then expressed in terms of a single field.

In general mixed and hybrid elements provide better accuracy than one-field models in both displacements and stresses, or the corresponding field variables, in cases other than solid mechanics. One of the most important applications of hybrid models is the treatment of stress and strain singularities. They have been extensively applied in the solution of two and three-dimensional crack problems. Hybrid models also have the advantage that they do not experience locking for elastic nearly incompressible materials and for the case of thin plates and shells with transverse shear effect.

1.4 Boundary Element Method Fundamentals

In contrast to the finite element method, which started by using physical considerations and whose mathematical foundations were only established some years later, the boundary element method has its origins on the theory of boundary integral equations. These have been applied to solve boundary-value problems of potential theory and classical elasticity since the beginning of this century.

The modern theory of boundary integral equations is assumed to begin with Fredholm. In his work [24], published in 1903, the existence of integral equation solutions was demonstrated through the use of a limiting discretization procedure.

From that date until the early sixties, the integral equation technique was used as an analytical method in the solution of boundary value problems, in potential theory and elastostatics. The work of Kellog [25], Muskhelishvili [26] and Kupradze [27] should be mentioned herein as important contributions in the area. However, due to difficulties in dealing with singular or weakly singular kernels that arise in integral equations and also, due to the lack of rigorous proofs of existence of solutions, the use of such technique was restricted to a limited number of problems. Furthermore, the technique of using integral equations as a tool for the solution of boundary-value problems was not well known among engineers at that time and its potential for obtaining numerical solutions was not exploited until the advent of computers.

In the sixties, as a consequence of the widespread use of the computer to solve problems of mathematical physics, some new and improved boundary integral formulations for potential theory and elasticity and also the corresponding numerical solutions were published. This fact can be considered the beginning of the boundary element method.

Some of these works formulated Fredholm integral equations of both first and second kinds, which come from the representation of a harmonic function by single layer or double layer potentials. Early papers presented with this methodology are the contributions from Hess and Smith [28], Jaswon and Ponter [29], Jaswon [30], Symm [31] and Massonnet [92]. This kind of formulation is the basis for the so-called indirect formulation.

Other formulations were based on either Green's formula for potential problems or in Somigliana's identity for elastostatics. These form the origin of the so-called direct formulation. The paper from Jaswon and Ponter [29], in 1963, appears to be the first published paper exploiting Green's formula on the boundary as a functional relationship between boundary potentials and its normal derivatives. The use of such relationship sets up the basis for the generation of direct formulations in the boundary element method.

Some years later, a similar formulation for elasticity problems was proposed by Rizzo [33], which used Somogliana's formula — the equivalent to Green's formula in elasticity — to generate his integral technique. Rizzo's formulation was solved numerically by Rizzo and Shippy [35] and Cruse [34]. Some other early and noticeable works employing the direct BEM are those from Cruse and Rizzo [36] and Cruse [37].

The indirect formulation uses unknown variables, without physical significance. The direct approach involves instead physical quantities, such as potentials and its normal derivatives (in the case of potential problems) or displacements and tractions (in the case of elasticity). This is one of the main reasons for the modern preference of the direct approach and for its wider acceptance by industry.

It was shown by Brebbia [40] that the direct boundary element method can also be formulated through weighted residual statements. The boundary element method was, therefore, shown to have a common basis with other numerical techniques, such as the finite element and finite difference methods. The equations for boundary elements, finite elements and finite differences can be generated through a collocation procedure, a Galerkin technique, and the method of moments, respectively. As a consequence of being formulated by different weighted residual approaches, these numerical methods have different features. As an example of such differences, boundary element formulations produce a fully-populated non-symmetric matrix, as opposed to the finite element method that produces a larger but banded symmetric matrix.

The boundary element method is considered to be a kind of mixed method by analogy with mixed finite element formulations: direct boundary element formulations usually involve two types of independent variables, which are boundary potential and its normal derivative, for potential problems and displacements and tractions, for elasticity.

1.5   Boundary Element Variational Formulations

To solve the boundary integral equations, the variational approach has been applied as an alternative to the collocation procedure, in both the direct and the indirect boundary element formulations. The Galerkin weighted residual approach has also been used as a discretization method. McDonald et al.

[43] and Jeng and Wexler [44, 45] have proposed numerical approaches with a variational basis for the solution of the boundary integral equations. It has been shown that the resulting discretized system of equations in [43] and in [45] are identical to those resulting from the Galerkin weighted residual method.

Modified Galerkin boundary formulations have also appeared in the literature. Hsiao, Kopp and Wendland [46] proposed the so-called Galerkin collocation method. Their approach demonstrated the existence of a solution of the algebraic system of equations, as well as the asymptotic convergence of the approximate solutions. References [47, 48, 49, 50] discuss this Galerkin-collocation method and the related error analysis and convergence proofs for different types of problems.

In elasticity, the Galerkin method was applied by Polizzotto [51] and Maier and Polizzotto [52] to derive a symmetric boundary element formulation. In Onishi [53], Galerkin's method is applied to the solution of the boundary integral equations in heat conduction; the convergence and stability of the approach is also proved.

The utilization of variational principles to generate the boundary integral equations is relatively recent. In the late eighties, some authors have developed boundary element formulations based on generalized principles such as the Hellinger-Reissner and the Wu-Washizu.

Dumont [54, 55] has proposed a hybrid stress boundary element formulation based on the Hellinger-Reissner principle. The formulation considers the stresses in the domain and the displacements on the boundary as independent functions. The resulting coefficient matrix is symmetric. The classical point force fundamental solution is used in this formulation to approximate the domain variable.

DeFigueiredo and Brebbia [56] presented an assumed-displacement hybrid boundary formulation, for the solution of potential problems. It is based on a generalized variational principle, which involve three independent boundary variables, two of them defined on the boundary and the other in the domain. The approach uses the same type of approximation for the domain variable applied in [54, 55] . This hybrid boundary formulation was later extended to elastostatic problems [57].

Polizzotto [58] used either the Hu-Washizu principle or the Hellinger-Reissner principle to derive a boundary element formulation which leads to the same symmetric coefficient matrix found by the Galerkin approach presented in [52]. He also proposed a boundary variational principle, which was called the boundary min-max principle. In this principle, the boundary displacements and tractions are considered as the independent field variables. The resulting boundary element formulation is shown to be equivalent to the previous Galerkin approach, which shows symmetry and definiteness. This formulation has also been extended to include elastoplasticity [58] and elastodynamics [59].

Recently, Felippa [60] proposed a formulation for boundary elements in linear elastostatics, which is based on his parametrized displacement-generalized hybrid variational principle [42]. Distinct boundary element models can then be obtained by setting different values to a parameter. His approach can, therefore, be considered a systematic way of deriving boundary element formulations.

Chapter 2

Potential Problems

2.1 Introduction

Potential problems are here defined as those that can be expressed in terms of a potential function and are governed by either a Laplace or a Poisson equation. Sometimes these problems are more specifically referred to as scalar potential problems [61], as opposed to vector potential problems, which can be used to represent other physical phenomena such as those treated in the theory of elasticity.

Many steady-state field problems frequently encountered in engineering practice, such as heat conduction, distribution of electric and magnetic potential, seepage through porous media and irrotational flow of ideal fluids are governed by the same potential formulation. Some problems in solid mechanics can also be represented by a Laplace type potential. That is the case of torsion and bending of prismatic bars, problems that can be written in terms of a potential function. In torsion problems this function is called a warping function, while in bending problems it is known as a stress function.

Although these are very simple problems in comparison to others in mathematical physics, solutions for the governing equations cannot be obtained analytically for complicated geometries and boundary conditions. Because of this, numerical methods such as finite differences, finite elements and boundary elements have been applied successfully to the solution of this class of problems for many years.

In this chapter, a new boundary element formulation to solve potential problems is developed. It is based on a generalized variational principle involving three independent variables, two of them defined on the boundary and the other in the domain. The resulting model can therefore be classified as a hybrid one, in analogy with the hybrid finite element models [22, 23]. The variables defined on the boundary are the potential and its normal derivative, while another potential field is defined inside the domain.

In the proposed formulation, a domain integral can be reduced to an integral along the boundary. This is achieved by approximating the potential in the domain through the classical boundary element fundamental solution for potential problems [40, 64, 65]. Consequently, in this hybrid boundary element formulation for Laplace's equation, the final equations only involve boundary integrals. In the case of Poisson's equation there is only one volume integral appearing in the final expression. It is due to the non-homogeneous term and it is the same type of integral present in the classical boundary element method formulation of the problem [64, 65].

The other field variables, which are defined only on the boundary, are represented in terms of shape functions as it is normally done in the conventional boundary element method. The expression for the functional in matrix form can then be obtained. The final system of algebraic equations results from the stationary conditions of this functional.

An interesting feature of this formulation is that it leads to an equivalent stiffness approach or, in other words, the final equation involves a symmetric stiffness matrix and all the unknowns are potentials. Another important characteristic of the approach is that both the stiffness matrix and the unknown potentials are only defined on the boundary.

In section 2.3 the governing equations of the problem are presented. In section 2.4 the multi-field functional to be used is derived and its corresponding Euler equations (stationary conditions) are obtained. Section 2.5 is concerned with the derivation of the model. In section 2.6 the formulation is shown to be symmetric.

2.2   Indicial Notation

The indicial, or index notation, is a compact way of writing equations, especially when they involve summations and derivatives. It has been extensively employed in the literature for both elasticity and potential theories. It will be used throughout this work and therefore a brief review will be given herein.

In this notation the rectangular Cartesian coordinate axis, usually denoted by x, y, z, are instead called x_1, x_2, x_3. Thus the components of a vector $\mathbf{v}$ are represented as v_1, v_2, v_3 and similarly for the components of a tensor $\mathbf{T}$: T_{11}, T_{12}, T_{13}, T_{21} *etc.* A letter subscript is used to designate the coordinate axis. For instance, a component of $\mathbf{v}$ is denoted by v_i, where i may assume the values 1, 2 or 3.

One can easily verify that this is a compact notation. For example, the equation

$$a_i = b_i + c_i \tag{2.1}$$

in fact represents a set of three equations, in three dimensional domains, obtained by giving i the values 1, 2, or 3, successively.

Two further conventions are used within the index notation. The first is that *a repeated literal index in any term of an expression implies summation.* For example in three dimensions:

$$a_i a_i \; = \; a_1^2 + a_2^2 + a_3^2 \tag{2.2}$$

$$a_{kk} \; = \; a_{11} + a_{22} + a_{33} \tag{2.3}$$

$$a_{j2}\, b_{3j} \; = \; a_{12}\, b_{31} + a_{22}\, b_{32} + a_{32}\, b_{33} \tag{2.4}$$

This summation convention will not be used when the indices refer to an entry of a matrix, for instance, when specifying the main diagonal f_{ii} of a matrix $\mathbf{F}$.

The second convention is that *a comma preceding an index denotes partial differentiation with respect to the variable represented by that index.* Thus

$$u_{i,j} = \frac{\partial u_i}{\partial x_j} \tag{2.5}$$

Sometimes in this work this derivative convention will not be used, for the sake of clarity. However, there is no danger of confusion in the notation for such cases.

If the two conventions are combined, one has a powerful way of writing formulas in a short form. This makes them easier to remember and, undoubtedly, faster to read and write as, for example, in the expression $m_{i,i}$ whose meaning is

$$m_{i,i} = \frac{\partial m_1}{\partial x_1} + \frac{\partial m_2}{\partial x_2} + \frac{\partial m_3}{\partial x_3} \tag{2.6}$$

The Kronecker delta δ_{ij} is a useful symbol, often used in this notation. It is defined as follows:

$$\delta_{ij} = \begin{cases} 1 & \text{if } i = j, \\ 0 & \text{if } i \neq j. \end{cases} \tag{2.7}$$

In all the formulas in this work the indices are assumed to have a range of two in two dimensions and three in three dimensions, unless otherwise stated.

2.3 Basic Equations

In this section, the potential problem will be defined in terms of a set of governing equations associated with the problem's physical interpretation, rather then in terms of the classical Laplace or Poisson equations. It can be shown, however, that both set of equations are entirely equivalent. The characterization of the problem in this form is important when deriving the functional, and afterwards, when developing the formulation.

The formulation is equally applicable to all types of potential problems. The equations and variables presented below have different physical interpretations and meanings, depending on the problem under consideration.For simplicity, only the fluid flow case will be considered in what follows, as it is appropriate for our purposes and easy to generalize to other cases.

Let us consider the problem of a potential distribution in a domain Ω bounded by a closed surface Γ (figure 2.1).

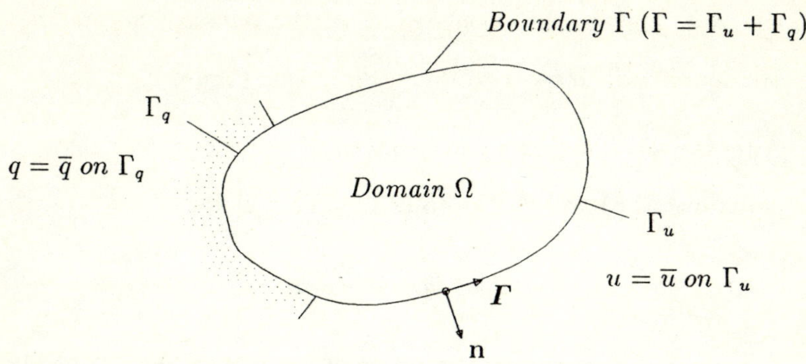

Figure 2.1: Domain and boundary conditions under consideration

The following equations are used to define the problem:

- An equation relating the potential u to the v_i components, which are in this case velocities, i.e.

$$v_i = \frac{\partial u}{\partial x_i} \qquad in \ \Omega \qquad (2.8)$$

- Another equation relating momentum density to velocity through the mass density ρ, which can be generally defined as a constitutive equation. This relationship is written as

$$m_i = \rho \, v_i \qquad in \ \Omega. \qquad (2.9)$$

- Equilibrium within the domain is satisfied by the following mass production equation

$$\frac{\partial m_i}{\partial x_i} + b = 0 \qquad in \ \Omega \qquad (2.10)$$

where b is the mass production density defined in Ω.

- If equation (2.10) is specialized at the boundary it gives the normal flux as

$$q = n_i\, m_i \qquad on \ \ \Gamma \tag{2.11}$$

where n_i are the direction cosines of the outward normal $\boldsymbol{n}$ to the boundary, with respect to the x_i directions (figure 2.1).

Alternatively the flux q can be given in terms of u by substitution of equations (2.8) and (2.9) into (2.11). This yields

$$q = \rho\, \frac{\partial u}{\partial x_i}\, n_i \tag{2.12}$$

It can also be represented in terms of the normal derivative of u as

$$q = \rho\, \frac{\partial u}{\partial n} \tag{2.13}$$

- The boundary conditions for potential and fluxes are as follows:

$$essential \ \ conditions: \ \ u = \overline{u} \qquad on \ \ \Gamma_u \tag{2.14}$$

$$natural \ \ conditions: \ \ q = \overline{q} \qquad on \ \ \Gamma_q \tag{2.15}$$

where $\Gamma = \Gamma_u + \Gamma_q$ (see figure 2.1).

It will now be shown that, when ρ is constant, the set of equations (2.8), (2.9) and (2.10) is equivalent to Laplace's equation or to Poisson's equation depending on b being equal to zero or otherwise.

Consider the relationship between m_i and u by substituting v_i from equation (2.8) into (2.9) as shown:

$$m_i = \rho\, \frac{\partial u}{\partial x_i} \qquad in \ \ \Omega \tag{2.16}$$

By taking into consideration expression (2.16), equation (2.10) is transformed into the following form:

$$\frac{\partial}{\partial x_i}\left(\rho\,\frac{\partial u}{\partial x_i}\right) + b = 0 \tag{2.17}$$

If ρ is constant, one obtains Poisson's equation which is written below

$$\nabla^2 u = -\frac{b}{\rho} \tag{2.18}$$

where $\nabla^2(\ )$ is the Laplace operator defined as

$$\nabla^2(\ ) = \frac{\partial^2(\ )}{\partial x_i^2} \tag{2.19}$$

Equation (2.18) can also be written as

$$\nabla^2 u = f \tag{2.20}$$

where

$$f = -\frac{b}{\rho} \tag{2.21}$$

When there is no mass production inside the domain (i.e. $b = 0$) Laplace's equation is obtained:

$$\nabla^2 u = 0 \tag{2.22}$$

From these results one can conclude that equations (2.8), (2.9) and (2.10) are equivalent to equation (2.20), or (2.22) when $b = 0$. Hence any of these two sets of equations, together with equation (2.11) and boundary conditions (2.14) and (2.15), can be used to define the potential problem.

2.4   Generalized Variational Principle

The classical one-field variational principle for the solution of the potential problem defined in the previous section is a minimum principle and can be stated as follows [62, 63]:

The solution of the potential problem is the function u which minimizes the functional $\mathcal{I}(u)$ given by

$$\mathcal{I}(u) \;=\; \int_{\Omega} \left(\frac{1}{2}\, \rho\, \frac{\partial u}{\partial x_i}\, \frac{\partial u}{\partial x_i} - b\, u \right)\, d\Omega \;-\; \int_{\Gamma_q} \overline{q}\, u\, d\Gamma \qquad (2.23)$$

with the boundary condition (2.14) (repeated here for convenience):

$$u = \overline{u} \qquad \text{on } \Gamma_u \qquad (2.24)$$

By assuming that equations (2.8), (2.9) and (2.11) are identically satisfied it is possible to write expression (2.23) as

$$\mathcal{I}(u) \;=\; \int_{\Omega} \left(\frac{1}{2}\, m_i\, v_i - b\, u \right)\, d\Omega \;-\; \int_{\Gamma_q} \overline{q}\, u\, d\Gamma \qquad (2.25)$$

where m_i and v_i are functions of u.

If one wishes the potential in the domain — u — and the potential on the boundary — $\tilde{u}$ — to be independent field variables, the compatibility condition (2.27) between these two fields on the boundary needs to be introduced as a subsidiary condition. The variational form of the problem is then

$$\mathcal{I}_1\,(u, \tilde{u}) \;=\; \int_{\Omega} \left(\frac{1}{2}\, m_i\, v_i - b\, u \right)\, d\Omega \;-\; \int_{\Gamma_q} \overline{q}\, \tilde{u}\, d\Gamma \qquad (2.26)$$

with the subsidiary compatibility condition:

$$u = \tilde{u} \qquad \text{on } \Gamma \qquad (2.27)$$

and the essential boundary condition given by

$$\tilde{u} = \overline{u} \qquad \text{on } \Gamma_u \qquad (2.28)$$

The subsidiary condition (2.27) can be introduced into the variational expression by using a Lagrange multiplier, λ. The modified variational principle can now be stated as:

The solution of the problem is given by the satisfaction of the stationary conditions of the multi-field functional $\mathcal{I}_2$, defined by

$$\mathcal{I}_2\,(u,\tilde{u},\lambda) \;=\; \int_\Omega \left(\frac{1}{2}\,m_i\,v_i - b\,u\right)\,d\Omega - \int_{\Gamma_q} \bar{q}\,\tilde{u}\,d\Gamma + \int_\Gamma \lambda\,(\tilde{u}-u)\,d\Gamma \qquad (2.29)$$

with the boundary condition:

$$\tilde{u} \;=\; \overline{u} \qquad \text{on } \Gamma_u \qquad\qquad (2.30)$$

Notice that the new functional $\mathcal{I}_2$ has three independent field variables and the modified principle is no longer a *minimum principle* but a *stationary principle.*

The Euler equations can then be obtained as the stationary conditions of $\mathcal{I}_2$. The first variation of the functional $\mathcal{I}_2$ is determined by taking variations with respect to the three independent variables in equation (2.29), namely u, $\tilde{u}$ and λ. It can therefore be written as follows:

$$\delta\mathcal{I}_2(u,\tilde{u},\lambda) \;=\; \int_\Omega \left[\frac{1}{2}(m_i\,\delta v_i + v_i\,\delta m_i) - b\,\delta u\right]\,d\Omega - \int_{\Gamma_q} \bar{q}\,\delta\tilde{u}\,d\Gamma$$

$$\int_\Gamma \lambda\,\delta\tilde{u}\,d\Gamma - \int_\Gamma \lambda\,\delta u\,d\Gamma + \int_\Gamma (\tilde{u}-u)\,\delta\lambda\,d\Gamma \qquad (2.31)$$

By taking into account that ρ is constant and that

$$\frac{\partial}{\partial x_i}(\delta u) = \delta\left(\frac{\partial u}{\partial x_i}\right) \qquad\qquad (2.32)$$

after applying equations (2.8) and (2.9), one can write the following:

$$v_i\,\delta m_i = \dot{m}_i\,\delta v_i = m_i\,\frac{\partial}{\partial x_i}(\delta u) \qquad\qquad (2.33)$$

and

$$\frac{1}{2}(m_i\,\delta v_i + v_i\,\delta m_i) = m_i\,\frac{\partial}{\partial x_i}(\delta u) \qquad\qquad (2.34)$$

Assuming that the boundary condition $\tilde{u} = \bar{u}$ is satisfied on Γ_u , i.e. $\delta\tilde{u} = 0$ on Γ_u , and substituting (2.34) into (2.31) yields:

$$\delta\mathcal{I}_2\,(u,\tilde{u},\lambda) \;=\; \int_\Omega \left[m_i\,\frac{\partial}{\partial x_i}(\delta u) - b\,\delta u \right]\,d\Omega + \int_{\Gamma_q} (\lambda - \bar{q})\,\delta\tilde{u}\,d\Gamma \;+$$
$$\int_\Gamma \lambda\,\delta u\,d\Gamma \;+\; \int_\Gamma (\tilde{u} - u)\,\delta\lambda\,d\Gamma \qquad (2.35)$$

It is possible to apply Green's first identity [66] to transform the first term in the domain integral as shown:

$$\int_\Omega m_i\,\frac{\partial}{x_i}\,(\delta u)\,d\Omega = \int_\Gamma m_i\,n_i\,\delta u\,d\Gamma - \int_\Omega \frac{\partial m_i}{\partial x_i}\,\delta u\,d\Omega \qquad (2.36)$$

The substitution of the relationship between flux and potential (2.11) into equation (2.36) leads to

$$\int_\Omega m_i\,\frac{\partial}{\partial x_i}\,(\delta u)\,d\Omega = \int_\Gamma q\,\delta u\,d\Gamma - \int_\Omega \frac{\partial m_i}{\partial x_i}\,\delta u\,d\Omega \qquad (2.37)$$

One should notice that the variables q and m_i are functions of the independent variable u.

By taking equation (2.37) into the expression for the first variation of $\mathcal{I}_2$, as given by equation (2.35), and rearranging the terms, the following expression is obtained:

$$\delta\mathcal{I}_2\,(u,\tilde{u},\lambda) \;=\; -\int_\Omega \left(\frac{\partial m_i}{\partial x_i} + b \right)\,\delta u\,d\Omega \;+\; \int_\Gamma (\tilde{u} - u)\,\delta\lambda\,d\Gamma \;+$$
$$\int_\Gamma (q - \lambda)\,\delta u\,d\Gamma \;+\; \int_{\Gamma_q} (\lambda - \bar{q})\,\delta\tilde{u}\,d\Gamma \qquad (2.38)$$

This first variation, equation (2.38), must vanish for any values of δu , $\delta\tilde{u}$ and $\delta\lambda$ and hence the Euler equations for the functional are the following:

$$\frac{\partial m_i}{\partial x_i} + b \;= 0 \qquad\qquad \text{in } \Omega \qquad\qquad (2.39)$$

$$\tilde{u} - u \;= 0 \qquad\qquad \text{on } \Gamma \qquad\qquad (2.40)$$

$$q - \lambda \;= 0 \qquad\qquad \text{on } \Gamma \qquad\qquad (2.41)$$

$$\lambda - \bar{q} \;= 0 \qquad\qquad \text{on } \Gamma \qquad\qquad (2.42)$$

The physical meaning of the Lagrange multiplier is determined by the last two equations, (2.41) and (2.42), in other words, the Lagrange multiplier λ is equal to the normal flux defined on the boundary, which will now be called $\tilde{q}$, to differentiate it from the internal flux q given by equation (2.13). It is seen that the Euler equations (2.39) to (2.42), together with the a priori assumed satisfied equations (2.8), (2.9), (2.11) and (2.30), completely define the potential problem.

If one now substitutes the Lagrange multiplier by the corresponding physical variable — $\tilde{q}$ — the functional $\mathcal{I}_2$ in (2.29) can be recast into the following form:

$$\mathcal{I}_2\left(u,\tilde{u},\tilde{q}\right) = \int_\Omega \left(\frac{1}{2}\,m_i\,v_i - b\,u\right)\,d\Omega + \int_\Gamma \tilde{q}\,(\tilde{u}-u)\,d\Gamma - \int_{\Gamma_q} \bar{q}\,u\,d\Gamma \quad (2.43)$$

Alternatively taking into account expression (2.8) one can write

$$\mathcal{I}_2\left(u,\tilde{u},\tilde{q}\right) = \int_\Omega \left(\frac{1}{2}\,m_i\,\frac{\partial u}{\partial x_i} - b\,u\right)\,d\Omega + \int_\Gamma \tilde{q}\,(\tilde{u}-u)\,d\Gamma - \int_{\Gamma_q} \bar{q}\,u\,d\Gamma$$
$$(2.44)$$

Green's first identity [66] can be used to transform the first term of the right-hand-side in equation (2.44) as shown below:

$$\int_\Omega \frac{1}{2}\,m_i\,\frac{\partial u}{\partial x_i}\,d\Omega = \int_\Gamma \frac{1}{2}\,u\,q\,d\Gamma - \int_\Omega \frac{1}{2}\,u\,\frac{\partial m_i}{\partial x_i}\,d\Omega \quad (2.45)$$

Finally, by substituting the results from (2.45) into equation (2.44), the functional $\mathcal{I}_{HP}$ can be written as

$$\mathcal{I}_{HP}\left(u,\tilde{u},\tilde{q}\right) = \int_\Gamma \frac{1}{2}\,u\,q\,d\Gamma - \int_\Gamma \tilde{q}\,(u-\tilde{u})\,d\Gamma - \int_\Gamma \bar{q}\,\tilde{u}\,d\Gamma -$$
$$\int_\Omega b\,u\,d\Omega - \int_\Omega \frac{1}{2}\,u\,\frac{\partial m_i}{\partial x_i}\,d\Omega \quad (2.46)$$

where

$\mathcal{I}_{HP}$ is a hybrid functional which depends on u, $\tilde{u}$ and $\tilde{q}$,

u is the potential defined in the domain,

$\tilde{u}$ and $\tilde{q}$ are, respectively, potential and flux, both defined on the boundary,

$\overline{q}$ is the prescribed value of normal flux valid on Γ_q,

$i = 1, 2$ for two-dimensional domains and $i = 1, 2, 3$ in three dimensions. Einstein's summation convention for repeated indices is implied.

Expression (2.46) gives the final form of the multi-field functional and is the starting point for the model proposed in this work.

The solution of the problem is given by the functions which make stationary the functional $\mathcal{I}_{HP}$, in equation (2.46). The admissible functions are defined as those u which satisfy (2.8) and (2.11), $\tilde{u}$ which fulfills the essential boundary condition (2.30) and $\tilde{q}$, piecewise continuous functions on the boundary.

2.5 Derivation of the Model

In this section, the hybrid displacement boundary element formulation is obtained by representing the three independent field variables — u, $\tilde{u}$ and $\tilde{q}$ — through approximate functions and then applying the generalized variational principle described in the previous section. The stationary conditions for the functional obtained give an approximate solution to the problem. The quality of this solution depends on how precisely the actual solution can be represented.

2.5.1 Definition of Fundamental Solution

Before describing how the variables are approximated a review of some definitions used in the classical boundary element method [40, 64, 65] and adopted in this work is presented.

Fundamental solution to Laplace's equation:

In an infinite domain, the value of the potential at a point x, due to a *concentrated unit source* at a point ξ, is called the *fundamental solution* at point x due to a source at ξ. It is usually denoted by $u^*(\xi, x)$. The point x, is called the *field point*. The point ξ, where the concentrated unit source is applied, is defined as the *source point*.

Source:

The source is the entity that generates the potential field. It varies according to the physical problem under consideration. For example; in problems concerning gravitational, electric or temperature fields the source is a mass, an electric charge or a heat source.

The fundamental solution for Laplace's equation satisfies the equation:

$$\nabla^2 u^*(\xi, x) = -\Delta(\xi, x) \tag{2.47}$$

where $\Delta(\xi, x)$ is the Dirac delta function which has the following property:

$$\int_\Omega u(x)\, \Delta(\xi, x)\, d\Omega = u(\xi) \tag{2.48}$$

This fundamental solution corresponds to the Newtonian potential for three-dimensional problems or the logarithmic potential for two-dimensional problems. In isotropic domains they are given by

$$u^* = \frac{1}{2\pi}\, log\left(\frac{1}{r}\right) \qquad \text{for 2D problems} \tag{2.49}$$

and

$$u^* = \frac{1}{4\pi\,r} \qquad \text{for 3D problems} \qquad (2.50)$$

where r is the distance between the points x and ξ.

These fundamental solutions are singular when the points x and ξ coincide, i. e. their values tend to infinity at the source point. They are the only fundamental solutions which enter in the formulation.

2.5.2 Approximation for the Domain Variable

Herein, and in the next four subsections, the matrix notation instead of the indicial notation will be used. In this manner, the final system of equations is obtained in matrix form which is simple and convenient.

The potential field at a point x inside the domain Ω — $u(x)$ — will be approximated as a series of products of fundamental solutions and unknown parameters. It can be represented as follows:

$$u(x) = \mathbf{u}^{*T}\boldsymbol{\gamma} \qquad x \ \epsilon \ \Omega. \qquad (2.51)$$

where $\mathbf{u}^*$ and $\boldsymbol{\gamma}$ are vectors given by

$$\mathbf{u}^* = \left\{ \begin{array}{c} u^{*1} \\ u^{*2} \\ \vdots \\ u^{*N} \end{array} \right\} \qquad (2.52)$$

and

$$\boldsymbol{\gamma} = \left\{ \begin{array}{c} \gamma_1 \\ \gamma_2 \\ \vdots \\ \gamma_N \end{array} \right\} \qquad (2.53)$$

in which N is the number of boundary points to be chosen in each particular case. These correspond to the points where the boundary variables will be evaluated.

The components of both vectors $\mathbf{u}$ and $\boldsymbol{\gamma}$ can be defined if we imagine that there is an infinite domain Ω_∞ that encompasses and has the same properties as Ω. The relative position between the points x and i in Ω_∞ must be the same as in the original domain Ω.

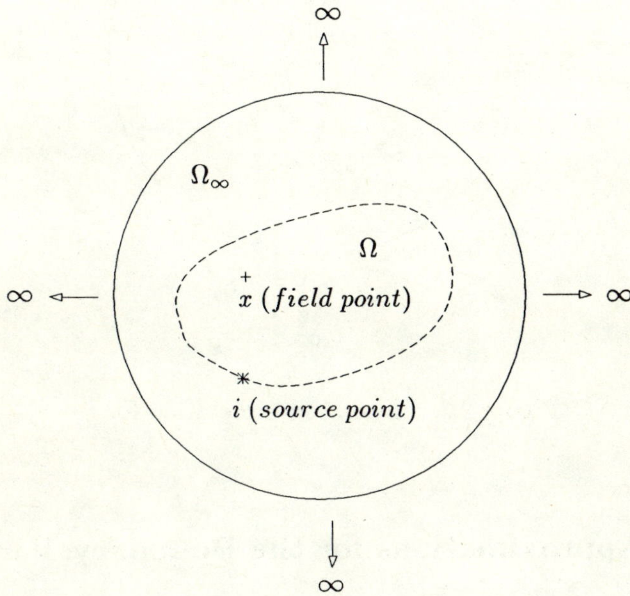

Figure 2.2: Domain Ω_∞

A component of vector $\mathbf{u}^*$, here written as u^{*i} ($i = 1, 2, ..., N$), is the fundamental solution at point x due to a concentrated unit source at point i (see figure 2.2). To be consistent with the notation adopted in section (2.5.1) it should be called $u^*(i, x)$. The short terminology is favoured here in order to achieve a more compact notation, which will be important in future sections.

The γ_i components ($i = 1, 2, ..., N$) can be viewed as fictitious concentrated sources, applied in this infinite domain at points which correspond to the boundary points i.

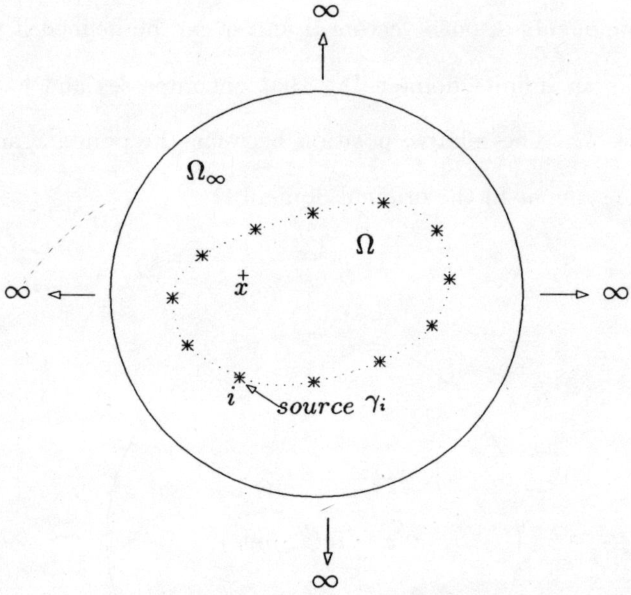

Figure 2.3: Physical interpretation for γ_i

2.5.3   Approximations for the Boundary Variables

The field variables defined on the boundary are herein represented with a tilde, $\tilde{u}$ and $\tilde{q}$. They will be approximated in the way which is normally adopted in finite or boundary elements. They are therefore approximated as the product of known interpolation functions and unknown parameters. The boundary potential and the normal flux on the boundary can then be written as

$$\tilde{u} = \mathbf{\Phi}^T \mathbf{u} \qquad on \ \ \Gamma \qquad (2.54)$$

$$\tilde{q} = \mathbf{\Psi}^T \mathbf{q} \qquad on \ \ \Gamma \qquad (2.55)$$

where $\boldsymbol{\Phi}$ and $\boldsymbol{\Psi}$ are matrices whose terms are interpolation functions of the type used in finite or boundary elements. T denotes transpose; $\mathbf{u}$ and $\mathbf{q}$ are vectors and their components are nodal values for the boundary potential $\tilde{u}$ and the normal flux $\tilde{q}$, respectively, i. e.

$$\mathbf{u} = \left\{ \begin{array}{c} u_1 \\ u_2 \\ \vdots \\ u_N \end{array} \right\} \tag{2.56}$$

$$\mathbf{q} = \left\{ \begin{array}{c} q_1 \\ q_2 \\ \vdots \\ q_N \end{array} \right\} \tag{2.57}$$

where u_i and q_i $(i = 1, 2, ..., N)$ are respectively the potential and the normal flux at a boundary node i. N is the number of boundary nodes considered.

Notice that the vectors u and q represent, respectively, the nodal values of potential and fluxes on the boundary although, for simplicity, the tilde has been ommited from the notation.

The boundary nodes are the same boundary points i where the sources γ_i (section 2.5.2) are applied.

2.5.4   Final System of Equations

The next step in the derivation of the formulation is to introduce the approximate functions u, $\tilde{u}$ and $\tilde{q}$ into the functional (2.46).

As mentioned before, the functions u^{*i} are singular at the boundary nodes i. To avoid introducing singular functions into the variational functional, these singularities are excluded from the domain under consideration. A new domain Ω' is then defined by removing from Ω parts of small spheres of radius ε, centered at the boundary nodes i $(i = 1, 2, \ldots, N)$, as shown in figure 2.4. The original domain Ω can be recovered from Ω' by taking the limit when the spheres' radius tends to zero, i. e.

when $\varepsilon \longrightarrow 0$ then

$$\Omega' \longrightarrow \Omega, \qquad \Gamma' \longrightarrow \Gamma, \qquad \Gamma'_u \longrightarrow \Gamma_u, \qquad \Gamma'_q \longrightarrow \Gamma_q$$

Γ', Γ'_u and Γ'_q are the boundaries of Ω' corresponding to the respective boundaries, which have already been defined for Ω and written without the prime (figure 2.4).

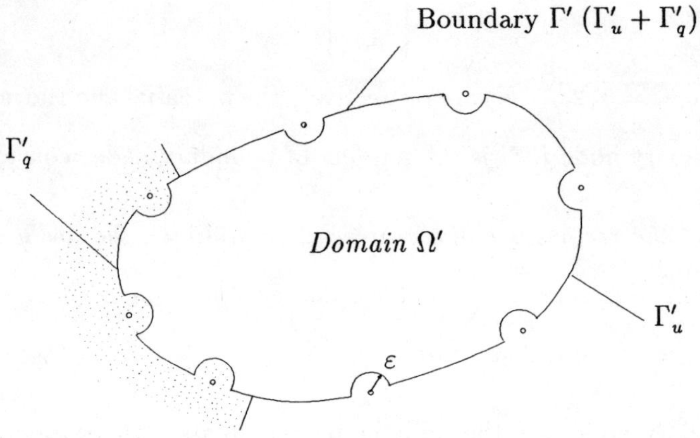

Figure 2.4: Definition of the domain Ω' and boundary Γ'

The functional (2.46) taken to the limit can now be expressed as

$$\mathcal{I}_{HP}(u, \tilde{u}, \tilde{q}) = \lim_{\varepsilon \to 0} \left\{ \int_{\Gamma'} \frac{1}{2}\, u\, q\, d\Gamma - \int_{\Gamma'} \tilde{q}\,(u - \tilde{u})\, d\Gamma - \int_{\Gamma'_q} \bar{q}\, \tilde{u}\, d\Gamma - \right.$$
$$\left. \int_{\Omega'} b\, u\, d\Omega - \int_{\Omega'} \frac{1}{2}\, u\, \frac{\partial m_k}{\partial x_k}\, d\Omega \right\} \tag{2.58}$$

(with $k = 1, 2$ in two-dimensions and $k = 1, 2, 3$ in three-dimensions)

One can now introduce the approximate variables — u, $\tilde{u}$ and $\tilde{q}$ — from equations (2.51), (2.54) and (2.55) into (2.58). The functional is then written in matrix notation as

$$\mathcal{I}_{HP} = \lim_{\varepsilon \to 0} \left\{ \int_{\Gamma'} \frac{1}{2}\, \mathbf{u}^{*T} \boldsymbol{\gamma}\, \mathbf{q}^{*T} \boldsymbol{\gamma}\, d\Gamma - \int_{\Gamma'} \boldsymbol{\Psi}^{T} \mathbf{q}\, (\mathbf{u}^{*T}\boldsymbol{\gamma} - \boldsymbol{\Phi}^{T}\mathbf{u})\, d\Gamma - \right.$$
$$\left. \int_{\Gamma'_q} \bar{q}\, \boldsymbol{\Phi}^{T}\mathbf{u}\, d\Gamma - \int_{\Omega'} b\, \mathbf{u}^{*T}\boldsymbol{\gamma}\, d\Omega \right\} \tag{2.59}$$

where $\mathbf{q}^*$ is the vector of fundamental fluxes associated with the vector of fundamental solutions $\mathbf{u}^*$. It is therefore expressed by equation (2.13). The domain Ω and the infinite domain Ω_∞ have the same properties, and consequently, the same mass density ρ. So vector $\mathbf{q}^*$ can be expressed by the following formula:

$$\mathbf{q}^* = \rho\, \frac{\partial \mathbf{u}^*}{\partial n} \tag{2.60}$$

The integral

$$\int_{\Omega'} \frac{1}{2}\, u\, \frac{\partial m_k}{\partial x_k}\, d\Omega \tag{2.61}$$

in equation (2.58) vanishes in consequence of the kind of approximation used for the potential in the domain. This will be demonstrated in what follows:

The fact that the potential in the domain is approximated by (2.51) implies that

$$m_k = \mathbf{m}_k^{*T} \boldsymbol{\gamma} \tag{2.62}$$

where $\mathbf{m}_k^*$ corresponds to the vector of the fundamental solutions $\mathbf{u}^*$ (section 2.3).

The integral in (2.61), after introducing the approximate variables, becomes

$$\int_{\Omega'} \frac{1}{2} \, u \, \frac{\partial m_k}{\partial x_k} \, d\Omega = \int_{\Omega'} \frac{1}{2} \, \mathbf{u}^{*T} \, \boldsymbol{\gamma} \, \frac{\partial \mathbf{m}_k^{*T}}{\partial x_k} \, \boldsymbol{\gamma} \, d\Omega \qquad (2.63)$$

According to equation (2.47), the fundamental solution satisfies Laplace's equation in the new domain Ω' and also satisfies equation (2.10) (see section 2.3). Hence one can write

$$\frac{\partial \mathbf{m}_k^*}{\partial x_k} = \mathbf{0} \qquad \text{in } \Omega' \qquad (2.64)$$

Substituting the above result into the expression (2.63), yields

$$\int_{\Omega'} \frac{1}{2} \, u \, \frac{\partial m_k}{\partial x_k} \, d\Omega = 0 \qquad (2.65)$$

One have then proven that the last integral in equation (2.58) vanishes.

It is possible to rearrange the product of matrices and vectors in equation (2.59). In addition to that, if the constant terms are taken out of the integral sign, the functional $\mathcal{I}_{HP}$ can be expressed as

$$\mathcal{I}_{HP} = \lim_{\varepsilon \to 0} \left\{ \frac{1}{2} \, \boldsymbol{\gamma}^T \left(\int_{\Gamma'} \mathbf{u}^* \, \mathbf{q}^{*T} \, d\Gamma \right) \boldsymbol{\gamma} - \mathbf{q}^T \left(\int_{\Gamma'} \boldsymbol{\Psi} \, \mathbf{u}^{*T} \, d\Gamma \right) \boldsymbol{\gamma} + \right.$$
$$\left. \mathbf{q}^T \left(\int_{\Gamma'} \boldsymbol{\Psi} \, \boldsymbol{\Phi}^T \, d\Gamma \right) \mathbf{u} - \mathbf{u}^T \left(\int_{\Gamma_q'} \bar{q} \, \boldsymbol{\Phi} \, d\Gamma \right) - \boldsymbol{\gamma}^T \left(\int_{\Omega} b \, \mathbf{u}^* \, d\Omega \right) \right\}$$

$$(2.66)$$

The following matrices can now be defined:

$$\mathbf{F} = \lim_{\varepsilon \to 0} \int_{\Gamma'} \mathbf{u}^* \, \mathbf{q}^{*T} \, d\Gamma \qquad (2.67)$$

$$\mathbf{G} = \lim_{\varepsilon \to 0} \int_{\Gamma'} \mathbf{u}^* \, \boldsymbol{\Psi}^T \, d\Gamma \qquad (2.68)$$

$$\mathbf{L} = \lim_{\varepsilon \to 0} \int_{\Gamma'} \boldsymbol{\Psi} \, \boldsymbol{\Phi}^T \, d\Gamma = \int_{\Gamma} \boldsymbol{\Psi} \, \boldsymbol{\Phi}^T \, d\Gamma \qquad (2.69)$$

$$\overline{\mathbf{Q}} = \lim_{\varepsilon \to 0} \int_{\Gamma_q'} \bar{q} \, \boldsymbol{\Phi} \, d\Gamma = \int_{\Gamma_q} \bar{q} \, \boldsymbol{\Phi} \, d\Gamma \qquad (2.70)$$

$$\mathbf{B} = \lim_{\varepsilon \to 0} \int_{\Omega'} b \, \mathbf{u}^* \, d\Gamma \qquad (2.71)$$

There are no singularities in matrices $\mathbf{L}$ and $\overline{\mathbf{Q}}$, consequently the limiting process is not necessary; they can be simply written in terms of the integral over the actual boundary, as indicated.

The substitution of these definitions into (2.66) allows the functional to be recast into a simpler form, as follows:

$$\mathcal{I}_{HP} = \frac{1}{2}\,\boldsymbol{\gamma}^T \mathbf{F}\,\boldsymbol{\gamma} - \mathbf{q}^T \mathbf{G}^T \boldsymbol{\gamma} + \mathbf{q}^T \mathbf{L}\,\mathbf{u} - \mathbf{u}^T \mathbf{Q} - \boldsymbol{\gamma}^T \mathbf{B} \qquad (2.72)$$

Stationary conditions for $\mathcal{I}_{HP}$ can now be found by taking variations in equation (2.72) with respect to u, $\tilde{u}$ and $\tilde{q}$; to be more precise, with respect to the unknown parameters — γ_i, u_i and q_i — which define those variables. This yields the following expression for the first variation of the functional $\mathcal{I}_{HP}$:

$$\begin{aligned}
\delta\mathcal{I}_{HP} &= \frac{1}{2}\,(\delta\boldsymbol{\gamma})^T \mathbf{F}\,\boldsymbol{\gamma} + \frac{1}{2}\,\boldsymbol{\gamma}^T \mathbf{F}\,\delta\boldsymbol{\gamma} - (\delta\mathbf{q})^T \mathbf{G}^T \boldsymbol{\gamma} - \mathbf{q}^T \mathbf{G}^T \delta\boldsymbol{\gamma} + \\
&\quad (\delta\mathbf{q})^T \mathbf{L}\,\mathbf{u} + \mathbf{q}^T \mathbf{L}\,\delta\mathbf{u} - (\delta\mathbf{u})^T \overline{\mathbf{Q}} - (\delta\boldsymbol{\gamma})^T \mathbf{B} \qquad (2.73)
\end{aligned}$$

The matrix $\mathbf{F}$ is symmetric, as will be proven in section 2.6. One can then use this property and rearrange the terms in (2.73) to obtain

$$\begin{aligned}
\delta\mathcal{I}_{HP} &= (\delta\boldsymbol{\gamma})^T \left(\mathbf{F}\,\boldsymbol{\gamma} - \mathbf{G}\,\mathbf{q} - \mathbf{B} \right) + \delta\mathbf{q}^T \left(-\mathbf{G}^T \boldsymbol{\gamma} + \mathbf{L}\,\mathbf{u} \right) + \\
&\quad (\delta\mathbf{u})^T \left(\mathbf{L}^T \mathbf{q} - \overline{\mathbf{Q}} \right) \qquad (2.74)
\end{aligned}$$

The functional $\mathcal{I}_{HP}$ is stationary when its first variation vanishes for any arbitrary values of $\delta\boldsymbol{\gamma}$, $\delta\mathbf{u}$ and $\delta\mathbf{q}$. Therefore the Euler equations or stationary conditions are

$$\mathbf{F}\,\boldsymbol{\gamma} - \mathbf{G}\,\mathbf{q} - \mathbf{B} = 0 \qquad (2.75)$$

$$-\mathbf{G}^T \boldsymbol{\gamma} + \mathbf{L}\,\mathbf{u} = 0 \qquad (2.76)$$

$$\mathbf{L}^T \mathbf{q} - \overline{\mathbf{Q}} = 0\,. \qquad (2.77)$$

The solution of the equations (2.75), (2.76) and (2.77) is the approximate solution of the problem as it gives the vectors γ , $\mathbf{u}$ and $\mathbf{q}$ which make the approximate functional stationary.

The way chosen to solve the above system of equation can vary. Herein it was preferred to express the unknowns γ and $\mathbf{q}$ in terms of $\mathbf{u}$. Then a final equation is obtained involving only the variable $\mathbf{u}$ as unknown, as will be shown in what follows.

$\mathbf{G}^T$ and $\mathbf{L}^T$ are non-singular square matrices and can then be inverted. Consequently, γ in equation (2.76) can be expressed in function of $\mathbf{u}$, as indicated below:

$$\gamma = \left(\mathbf{G}^T\right)^{-1} \mathbf{L}\,\mathbf{u} \tag{2.78}$$

and an expression for $\mathbf{q}$ in terms of $\mathbf{u}$ can also be found by substituting this result into equation (2.75), i. e.

$$\mathbf{q} = \mathbf{G}^{-1}\,\mathbf{F}\left(\mathbf{G}^T\right)^{-1}\mathbf{L}\,\mathbf{u} - \mathbf{G}^{-1}\,\mathbf{B} \tag{2.79}$$

In the case of Laplace's equation, the equilibrium requires that the integral of the fluxes along the boundary be zero and this is satisfied in a variational sense.

By replacing γ and $\mathbf{q}$ from equations (2.78) and (2.79) into (2.77) the resulting matrix equation can then be written as

$$\mathbf{K}\,\mathbf{u} - \mathbf{Q} = 0 \tag{2.80}$$

where

$$\mathbf{K} = \mathbf{R}^T\mathbf{F}\,\mathbf{R} \tag{2.81}$$

$$\mathbf{R} = \left(\mathbf{G}^T\right)^{-1}\mathbf{L} \tag{2.82}$$

$$\mathbf{Q} = \overline{\mathbf{Q}} + \mathbf{R}^T\mathbf{B} \tag{2.83}$$

Equation (2.80) shows that this hybrid boundary element formulation leads to an equivalent stiffness type approach: the matrix $\mathbf{K}$ is a hybrid boundary stiffness matrix which is symmetric, as it will be demonstrated in section 2.6; the vector $\mathbf{Q}$ is a consistent load vector.

Notice that matrix $\mathbf{K}$ is evaluated by carrying out integrations along the boundary only and that the unknown $\mathbf{u}$ is also calculated only on the boundary. The other primary unknowns — $\boldsymbol{\gamma}$ and $\mathbf{q}$ — can be obtained from the boundary potentials.

2.5.5 Solution on the Boundary

Boundary potentials:

The solution for the potential at boundary nodes — $\mathbf{u}$ — is obtained by solving equation (2.80).

Boundary fluxes:

Having the values for potentials, the normal fluxes at boundary nodes are evaluated using equation (2.79), i. e.

$$\mathbf{q} = \mathbf{G}^{-1}\,\mathbf{F}\,(\mathbf{G}^{T})^{-1}\,\mathbf{L}\,\mathbf{u} - \mathbf{G}^{-1}\,\mathbf{B} \tag{2.84}$$

2.5.6 Solution at Internal Points

Potentials at internal points:

Once the solution for potentials on the boundary is known, it is possible to calculate the vector $\boldsymbol{\gamma}$ from expression (2.78). The potentials at any internal point x inside the domain can then be computed simply using equation (2.51), repeated here for completeness:

$$u^{int}(x) = \mathbf{u}^{*T}\,\gamma \qquad x \,\epsilon\, \Omega \tag{2.85}$$

Fluxes at internal points:

Fluxes in x_ℓ direction at a point x in the domain — $q_\ell^{int}(x)$ — can be computed by differentiating equation (2.51) in x_ℓ direction, i. e.

$$\frac{\partial}{\partial x_\ell}\,[u^{int}(x)] = \frac{\partial \mathbf{u}^*}{\partial x_\ell}\,\gamma \tag{2.86}$$

and by substituting this result into (2.13). The following expression for the computation of internal fluxes is then found:

$$q_\ell^{int}(x) = \rho\,\frac{\partial \mathbf{u}^*}{\partial x_\ell}\,\gamma \tag{2.87}$$

2.6 Symmetry of the Stiffness Matrix

The hybrid boundary stiffness matrix $\mathbf{K}$ is given by equation (2.81), which is repeated below for convenience:

$$\mathbf{K} = \mathbf{R}^T\mathbf{F}\,\mathbf{R} \tag{2.88}$$

From that relationship it is possible to state that, if the matrix $\mathbf{F}$ is symmetric then the matrix $\mathbf{K}$ is also symmetric. Therefore the aim here is to prove the symmetry of $\mathbf{F}$.

Notice that matrix $\mathbf{F}$ was defined through equation (2.67) as

$$\mathbf{F} = \lim_{\epsilon \to 0} \int_{\Gamma'} \mathbf{u}^*\,\mathbf{q}^{*T}\,d\Gamma \tag{2.89}$$

with $\mathbf{u}^*$ and $\mathbf{q}^*$ given in sections 2.5.2 and 2.5.4, respectively.

By performing the product of the two vectors one finds the following matrix terms:

$$\mathbf{F} = \lim_{\varepsilon \to 0} \int_{\Gamma\prime} \begin{bmatrix} u^{*1}q^{*1} & u^{*1}q^{*2} & \ldots & u^{*1}q^{*N} \\ u^{*2}q^{*1} & u^{*2}q^{*2} & \ldots & u^{*2}q^{*N} \\ \vdots & \vdots & \vdots & \vdots \\ u^{*N}q^{*1} & u^{*N}q^{*2} & \ldots & u^{*N}q^{*N} \end{bmatrix} d\Gamma \qquad (2.90)$$

Consequently, an entry of the matrix $\mathbf{F}$ — f_{ij} — can be written as

$$f_{ij} = \lim_{\varepsilon \to 0} \int_{\Gamma\prime} u^{*i} \, q^{*j} \, d\Gamma(x) \qquad (2.91)$$

where u^{*i} is the fundamental solution at a field point x corresponding to a concentrated unit source at source point i $(i = 1, 2, ..., N.$ See section 2.5.2). Similarly q^{*j} is the fundamental normal flux at point x due to a concentrated unit source at j $(j = 1, 2, ..., N)$.

Notice that all the source and field points mentioned here are supposed to be located in the infinite domain, keeping the relative positions they would have if they were located on the boundary Γ of the actual domain Ω (figure 2.2).

Proof of the symmetry of $\mathbf{F}$:

Once all the boundary nodes (including the points i and j) have been excluded from the domain Ω' (figure 2.4), it is possible to write from equation (2.47) that

$$\nabla^2 u^{*i} = 0 \qquad \text{in } \Omega' \qquad (2.92)$$

and

$$\nabla^2 u^{*j} = 0 \qquad \text{in } \Omega' \qquad (2.93)$$

where u^{*j} is the fundamental solution corresponding to flux q^{*j}.

Multiplying equation (2.92) and (2.93) by u^{*j} and u^{*i}, respectively, gives

$$u^{*j} \ \nabla^2 u^{*i} = 0 \qquad \text{in } \Omega' \tag{2.94}$$

and

$$u^{*i} \ \nabla^2 u^{*j} = 0 \qquad \text{in } \Omega' \tag{2.95}$$

Equation (2.94) will now be subtracted from equation (2.95) and the result integrated over the domain Ω'. This yields

$$\int_{\Omega'} \left(u^{*i} \ \nabla^2 u^{*j} - u^{*j} \ \nabla^2 u^{*i} \right) d\Omega = 0 \tag{2.96}$$

The left-hand-side of the last equation can be transformed by applying Green's second identity [66] to obtain

$$\int_{\Gamma'} \left(u^{*i} \ \frac{\partial u^{*j}}{\partial n} - u^{*j} \ \frac{\partial u^{*i}}{\partial n} \right) d\Gamma = 0 \tag{2.97}$$

In section 2.3 it was assumed that ρ is constant. The above equation can then be multiplied by ρ and the result can be written as

$$\int_{\Gamma'} \left(u^{*i} \ \rho \ \frac{\partial u^{*j}}{\partial n} - u^{*j} \ \rho \ \frac{\partial u^{*i}}{\partial n} \right) d\Gamma = 0 \tag{2.98}$$

By applying equation (2.60) the previous equation can be transformed into the following form:

$$\int_{\Gamma'} u^{*i} \ q^{*j} \ d\Gamma = \int_{\Gamma'} u^{*j} \ q^{*i} \ d\Gamma \tag{2.99}$$

Now taking the limit on both sides of the previous equation, produces the following expression:

$$\lim_{\varepsilon \to 0} \int_{\Gamma'} u^{*i} \ q^{*j} \ d\Gamma = \lim_{\varepsilon \to 0} \int_{\Gamma'} u^{*j} \ q^{*i} \ d\Gamma \tag{2.100}$$

According to expression (2.91) one can finally write that

$$f_{ij} = f_{ji} \tag{2.101}$$

The symmetry of the flexibility type matrix $\mathbf{F}$ *has then been proved and so has the symmetry of the stiffness matrix* $\mathbf{K}$.

Chapter 3

Numerical Aspects in Potential Problems

3.1   Introduction

The procedures required for the numerical solution of the equations presented in the previous chapter will be explained in the following sections. Although only the two dimensional case will be considered throughout this chapter, similar considerations could be used for three dimensional problems.

A series of boundary integrals must be performed to obtain the matrices $\mathbf{F}$, $\mathbf{G}$ and $\mathbf{L}$, given by equations (2.67), (2.68) and (2.69) respectively. These matrices are necessary to obtain the system of equations (2.80) which produces the solution of the problem. To this end the boundary is subdivided into a number of elements over which the geometry is expressed in terms of appropriate algebraic functions and values at certain nodal points. Suitable numerical integration schemes are then used to evaluate these integrals.

To deal with the singular integrals which appear when evaluating the matrices $\mathbf{F}$ and $\mathbf{G}$, special techniques are applied either to calculate them analytically, if possible, or to transform these singular integrals in such a away that they can be integrated numerically.

The vector of equivalent nodal fluxes $\mathbf{Q}$, in the absence of internal sources, is obtained either directly (for constant elements) or through boundary integrations (for quadratic elements). The case of Poisson's equation, i. e. when internal sources are present, will be treated separately in section 3.4.

Two different types of elements will be developed in this chapter. For the first one of linear geometry, herein called constant element, the boundary variables — $\tilde{u}$ and $\tilde{q}$ — are assumed to have a constant value on each element. The second type of element is the isoparametric quadratic for which the geometry and the boundary variables — $\tilde{u}$ and $\tilde{q}$ — are interpolated through Lagrange polynomials, as it is usually done in the classical boundary element method [40, 64, 65].

For both elements the potential inside the domain — u — is approximated exactly in the same way, as it is indicated in equation (2.51). Consequently the matrices $\mathbf{F}$ for these two types of elements only differ because the boundary geometry is assumed to vary differently in each case. It is interesting to point out that the kind of approximation used for the boundary variables does not affect this matrix. The other matrices, $\mathbf{G}$ and $\mathbf{L}$, instead, vary with the type of approximations used for the geometry and for the boundary variables.

In this hybrid approach as in the conventional direct boundary element method, the same order of approximation for both $\tilde{u}$ and $\tilde{q}$ are used. This results in identical matrices $\boldsymbol{\Phi}$ and $\boldsymbol{\Psi}$. Therefore the matrix $\mathbf{G}$ in this hybrid formulation, defined by equation (2.68), is exactly the same matrix $\mathbf{G}$ that appears in the classical direct boundary element method formulation of the same problem [40, 64, 65].

3.2   The Constant Element

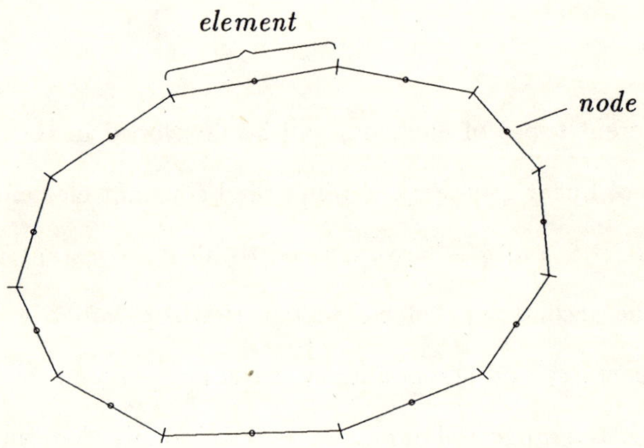

Figure 3.1: Discretization of the boundary by constant elements

For this type of element the boundary Γ is approximated by a series of straight elements, each of them associated with a node at its midpoint (figure 3.1). The boundary variables — the potential $\tilde{u}$ and the flux $\tilde{q}$ — are assumed to be constant on the element. This means that equations (2.54) and (2.55) can be written for an element Γ_k as:

$$\tilde{u} \; = \; u_k \qquad \text{on } \Gamma_k \tag{3.1}$$

$$\tilde{q} \;=\; q_k \qquad \text{on } \Gamma_k \qquad\qquad (3.2)$$

where k varies from 1 to N; for the constant element, the number of boundary elements is equal to the number of nodes, N.

The geometry of the element is linear and is defined by its extreme points, namely point (1) and point (2) as shown in figure 3.2.

The boundary integrations are then performed over each element. The integrals along the whole boundary Γ are evaluated by adding up the contributions of all the elements.

In order to compute the boundary integrals a homogeneous coordinate η is defined such that (figure 3.2):

$$\eta = \frac{2\,\Gamma}{\ell} \qquad\qquad (3.3)$$

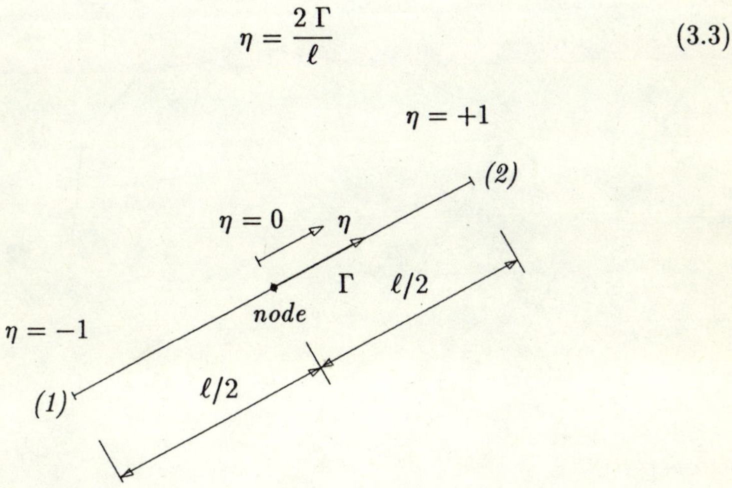

Figure 3.2: Homogeneous coordinate η

3.2.1 Matrix F for Constant Elements

The computation of matrix **F** is the most important operation in this approach because it involves integrating products of two singular functions, u^* and q^*. The terms of **F** are given by equation (2.91) which after discretization of the boundary can be written as:

$$f_{ij} = \lim_{\varepsilon \to 0} \sum_{k=1}^{N} \int_{\Gamma'_k} u^{*i}\, q^{*j}\, d\Gamma(x) \qquad i,j = 1,2,\ldots,N \qquad (3.4)$$

where ε was defined in section 2.5.4. Notice that u^{*i} represents the value of the fundamental solution at a point x when the source point is at i. The q^{*j} function defines the value of the fluxes of the fundamental solution at a point x corresponding to a source point at j.

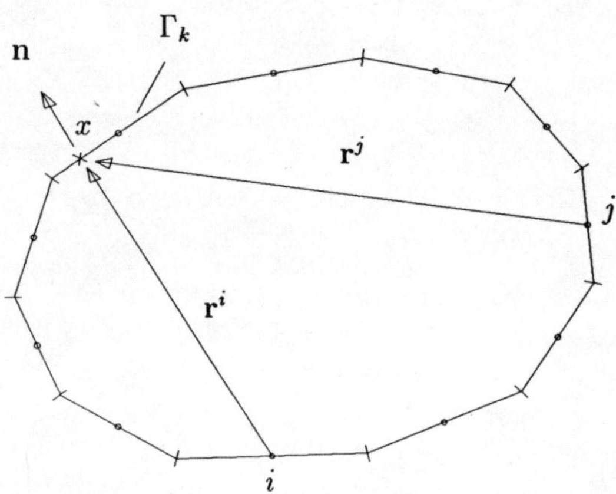

Figure 3.3: Sources i and j

According to equation (2.49) the fundamental solution u^{*i} for two-dimensional problems is given by:

$$u^{*i} = \frac{1}{2\pi}\, \ln\left(\frac{1}{r^i}\right) \qquad (3.5)$$

where r^i is the distance between the source point i and the field point x (figure 3.3)

From equation (2.60) the flux q^{*j} corresponding to the fundamental solution u^{*j} can be expressed as follows:

$$q^{*j} = \rho \, \frac{\partial u^{*j}}{\partial n} \tag{3.6}$$

where u^{*j} is given by

$$u^{*j} = \frac{1}{2\pi} \ln \left(\frac{1}{r^j} \right) \tag{3.7}$$

where r^j is the modulus of the vector $\mathbf{r}_j$ with origin at the source point and end at the field point x on Γ_k, as depicted in figure 3.3.

By substituting the expression for u^{*j} as given by equation (3.7) into equation (3.6), the flux q^{*j} can be given by

$$q^{*j} = - \frac{\rho}{2\pi} \frac{1}{r^j} \frac{\partial r^j}{\partial n} \tag{3.8}$$

where $\partial r^j / \partial n$ is the normal derivative of r^j with respect to the outward normal $\mathbf{n}$.

After taking equations (3.5) and (3.8) into expression (3.4) a generic coefficient f_{ij} becomes:

$$f_{ij} = - \frac{\rho}{4\pi^2} \lim_{\varepsilon \to 0} \sum_{k=1}^{n} \int_{\Gamma_k'} \ln \left(\frac{1}{r^i} \right) \frac{1}{r^j} \frac{\partial r^j}{\partial n} \, d\Gamma(x) \tag{3.9}$$

The evaluation of the integral in (3.9) over elements which include neither point i nor point j is not a difficult problem. It requires the computation of a regular integral on the element and hence it can be performed numerically by any suitable scheme. In the limit, however, when ϵ is equal to zero, this integral becomes singular when either the source point i of u^{*i} or the source point j of q^{*j} coincides with the field point x. It also becomes singular if both i and j coincide with x, as it happens for the elements in the main diagonal of $\mathbf{F}$.

Thus according to the type of singularity present in the integrals to be evaluated, an entry of the matrix $\mathbf{F}$ can be split into four different terms as follows:

$$f_{ij} = -\frac{\rho}{4\pi^2}\lim_{\varepsilon\to 0}\left\{\sum_{k=1,k\neq i,k\neq j}^{N}\int_{\Gamma_k'}\ln\left(\frac{1}{r^i}\right)\frac{1}{r^j}\frac{\partial r^j}{\partial n}\,d\Gamma + \int_{\Gamma_i'}\ln\left(\frac{1}{r^i}\right)\frac{1}{r^j}\frac{\partial r^j}{\partial n}\,d\Gamma + \right.$$
$$\left.\int_{\Gamma_j'}\ln\left(\frac{1}{r^i}\right)\frac{1}{r^j}\frac{\partial r^j}{\partial n}\,d\Gamma + \int_{\Gamma_i'}\ln\left(\frac{1}{r^i}\right)\frac{1}{r^i}\frac{\partial r^i}{\partial n}\,d\Gamma\right\} \qquad (3.10)$$

The boundaries Γ_k', Γ_i' and Γ_j' correspond to the original elements Γ_k, Γ_i and Γ_j in Ω where the singularities have been excluded. Their definition follows from the definition of the domain Ω' (section 2.5.4), as shown in figure 3.4.

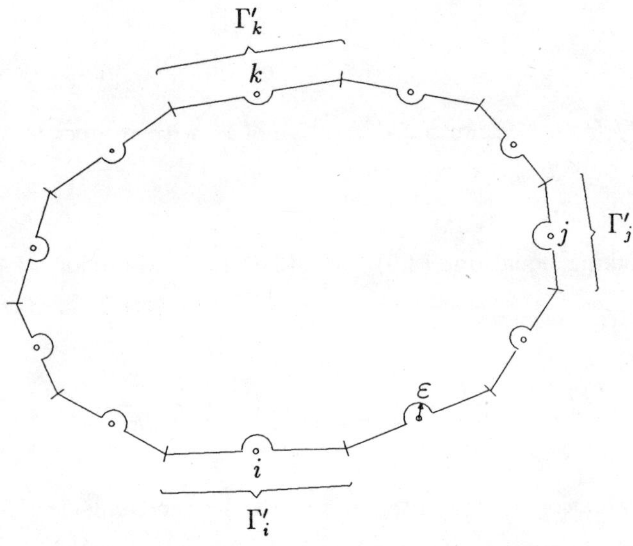

Figure 3.4: Domain Ω' for constant elements

In the first term on the right-hand-side of equation (3.10) the limit when ϵ tends to zero of the integral over Γ_k' is equal to the integral over Γ_k. This happens because the source points i and j are not on the element Γ_k and hence the functions u^{*i} and q^{*i} are continuous functions on that interval,

i. e. when the point x moves along the element. It is then possible to write f_{ij} as

$$f_{ij} = -\frac{\rho}{4\pi^2}\left\{ \sum_{k=1,k\neq i,k\neq j}^{N} \int_{\Gamma_k} \ln\left(\frac{1}{r^i}\right)\frac{1}{r^j}\frac{\partial r^j}{\partial n}\,d\Gamma + \lim_{\varepsilon\to 0}\int_{\Gamma_i'} \ln\left(\frac{1}{r^i}\right)\frac{1}{r^j}\frac{\partial r^j}{\partial n}\,d\Gamma + \right.$$

$$\left. \int_{\Gamma_j'} \ln\left(\frac{1}{r^i}\right)\frac{1}{r^j}\frac{\partial r^j}{\partial n}\,d\Gamma + \int_{\Gamma_i'} \ln\left(\frac{1}{r^i}\right)\frac{1}{r^i}\frac{\partial r^i}{\partial n}\,d\Gamma \right\} \qquad (3.11)$$

The four terms on the right-hand-side of the previous equation correspond to the four different types of integrals to be evaluated. Each one of these cases will be studied in detail in what follows.

Case 1: *Integration over an element which includes neither source point i nor source point j.*

The non-singular integral to be evaluated in this case will be denoted by I^k. According to equation (3.11) it is given by the following expression:

$$I^k = \int_{\Gamma_k} \ln\left(\frac{1}{r^i}\right)\frac{1}{r^j}\frac{\partial r^j}{\partial n}\,d\Gamma \qquad (3.12)$$

where r^i and r^j are shown in figure 3.3.

This integral can be computed numerically by using a Gauss-Legendre quadrature formula. One should be aware of the situation for which any of the sources i and j or both of them are close to the element over which the integration is being performed. In this case the integrand varies rapidly and sufficient integration points must be used to obtain the desired accuracy. It was found, by performing some tests, that six integration points are sufficient if the distances between the source points and the nearest point on the element are not greater than half the size of the element.

Case 2: *Integration over an element which includes the source point i.*

The integral to be computed corresponds to the second term in expression (3.11). It will be represented herein by I^i and can be written as:

$$I^i = \lim_{\varepsilon \to 0} \int_{\Gamma'_i} \ln\left(\frac{1}{r^i}\right) \frac{1}{r^j} \frac{\partial r^j}{\partial n}\, d\Gamma(x) \tag{3.13}$$

with the boundary Γ'_i and the vectors $\mathbf{r}^i$ and $\mathbf{r}^j$ shown in figure 3.5.

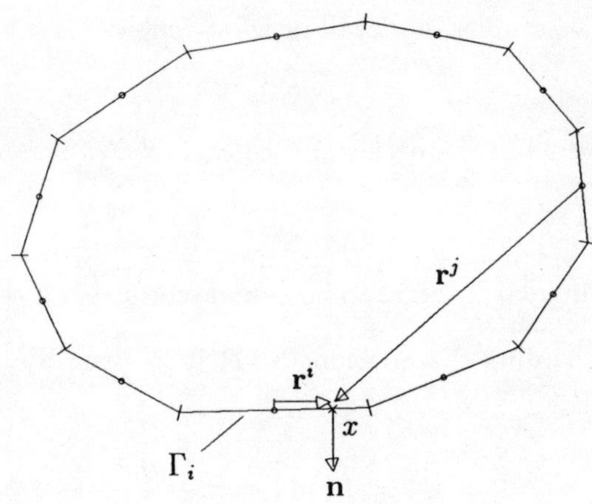

Figure 3.5: Integration over Γ_i

Equation (3.13) shows that in the limit the integral over Γ'_i contains a logarithmic type of singularity. Consequently the limit of the integral over Γ'_i is equal to the integral over Γ_i, since the latter exists and has a definite value and when $\varepsilon \to 0$ then $\Gamma'_i \to \Gamma_i$. Hence one can write

$$I^i = \int_{\Gamma_i} \ln\left(\frac{1}{r^i}\right) \frac{1}{r^j} \frac{\partial r^j}{\partial n}\, d\Gamma(x) \tag{3.14}$$

The function $(1/r^i)\,(\partial r^j/\partial n)$ is continuous over Γ_i and therefore the integral (3.14) can easily be evaluated through any suitable scheme such as a logarithmic Gaussian quadrature formula [67] or the scheme proposed by Telles [68], in which a transformation of variables is used to cancel the singularity.

Case 3: *Integration over an element which includes the source point j.*

Now the third term in equation (3.11) will be considered. The integral to be evaluated is then:

$$I^j = \lim_{\varepsilon \to 0} \int_{\Gamma'_j} \ln \left(\frac{1}{r^i} \right) \frac{1}{r^j} \frac{\partial r^j}{\partial n} \, d\Gamma(x) \qquad (3.15)$$

where r^i and r^j have been previously defined and the boundary Γ'_j, which corresponds to the boundary element Γ_j, is shown in figure 3.6.

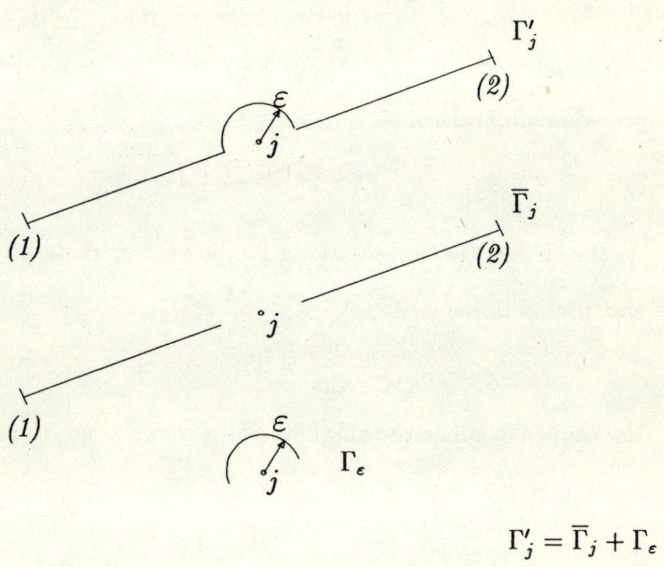

$$\Gamma'_j = \overline{\Gamma}_j + \Gamma_\varepsilon$$

Figure 3.6: Definition of Γ'_j

In order to evaluate the limit in expression (3.15), the boundary Γ'_j will be subdivided into two parts (figure 3.6): the first one corresponds to the straight part of Γ'_j, herein denoted by $\overline{\Gamma}_j$; the second one is the semi-circumference Γ_ε of radius ε.

Accordingly I^j can then be expressed as

$$I^j = \lim_{\varepsilon \to 0} \left\{ \int_{\overline{\Gamma}_j + \Gamma_\varepsilon} \ln\left(\frac{1}{r^i}\right) \frac{1}{r^j} \frac{\partial r^j}{\partial n} \, d\Gamma(x) \right\} \qquad (3.16)$$

and finally it can be written as follows:

$$I^j = \lim_{\varepsilon \to 0} \left\{ \int_{\overline{\Gamma}_j} \ln\left(\frac{1}{r^i}\right) \frac{1}{r^j} \frac{\partial r^j}{\partial n} \, d\Gamma(x) \right\} + \lim_{\varepsilon \to 0} \left\{ \int_{\Gamma_\varepsilon} \ln\left(\frac{1}{r^i}\right) \frac{1}{r^j} \frac{\partial r^j}{\partial n} \, d\Gamma(x) \right\}$$
$$(3.17)$$

If one uses the notation shown below:

$$\overline{I}^j = \lim_{\varepsilon \to 0} \left\{ \int_{\overline{\Gamma}_j} \ln\left(\frac{1}{r^i}\right) \frac{1}{r^j} \frac{\partial r^j}{\partial n} \, d\Gamma(x) \right\} \qquad (3.18)$$

$$I_\varepsilon = \lim_{\varepsilon \to 0} \left\{ \int_{\Gamma_\varepsilon} \ln\left(\frac{1}{r^i}\right) \frac{1}{r^j} \frac{\partial r^j}{\partial n} \, d\Gamma(x) \right\} \qquad (3.19)$$

the integral I^j can be evaluated by adding the results from the two previous integrals, i. e.

$$I^j = \overline{I}^j + I_\varepsilon \qquad (3.20)$$

As the element is straight along $\overline{\Gamma}^j$ the vectors $\mathbf{r}_j$ and $\mathbf{n}$ are orthogonal and $\mathbf{n}.\mathbf{r}^j = 0$, hence

$$\frac{\partial r^j}{\partial n} = 0 \qquad \text{on } \overline{\Gamma}_j \qquad (3.21)$$

Consequently since the singular point is excluded, the integral $\overline{I}^j$ vanishes, i. e.

$$\overline{I}^j = 0 \qquad (3.22)$$

The other integral I_ε will be computed analytically as follows:

From figure 3.7 one can see that over Γ_ε the following relationships hold:

$$r^j \;=\; \varepsilon \qquad (3.23)$$

$$\frac{\partial r^j}{\partial n} \;=\; \cos(\mathbf{r}_j, \mathbf{n}) = -1 \qquad (3.24)$$

$$d\Gamma \;=\; -\varepsilon \, d\theta \qquad (3.25)$$

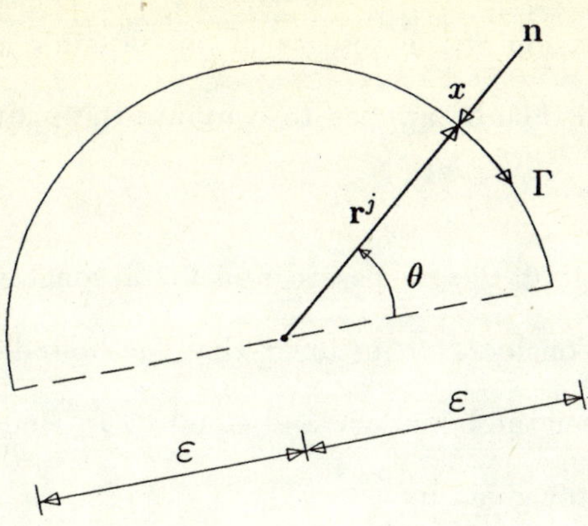

Figure 3.7: The boundary Γ_ε

Substituting these last expressions into equation (3.19) gives

$$I_\varepsilon = \lim_{\varepsilon \to 0} \int_\pi^0 \ln\left(\frac{1}{r^i}\right)\, d\theta \tag{3.26}$$

In this case, the limit of the integral is the integral of the limit and by performing the integral one has a final value for I_ε, as shown below:

$$I_\varepsilon = -\pi \ln\left(\frac{1}{R_{ij}}\right) \tag{3.27}$$

where R_{ij} is the distance between points i and j.

Case 4: *Integration over an element which includes both points i and j.*

This situation occurs when evaluating the main diagonal terms of $\mathbf{F}$, namely f_{ii}. The integral to be performed is the last term in equation (3.11), i. e.

$$I^{ii} = \lim_{\varepsilon \to 0} \int_{\Gamma_i'} \ln\left(\frac{1}{r^i}\right) \frac{1}{r^i} \frac{\partial r^i}{\partial n}\, d\Gamma(x) \tag{3.28}$$

Direct computation of this integral is not possible and physical considerations will then be applied to compute the main diagonal of the matrix $\mathbf{F}$.

The property that the fluxes are null for a constant potential field will be used. Consider, for instance, that the boundary potentials are constant all over the boundary and equal to 1. Hence $\mathbf{u} = \mathbf{I}$ where $\mathbf{I}$ is the unit vector given by

$$\mathbf{I} = \left\{ \begin{array}{c} 1 \\ 1 \\ \vdots \\ 1 \end{array} \right\} \tag{3.29}$$

Applying the property mentioned above, one can say that for $\mathbf{u} = \mathbf{I}$, $\mathbf{Q} = \mathbf{0}$. Taking these expressions into equation (2.80) one obtains

$$\mathbf{K}\,\mathbf{I} = \mathbf{0} \tag{3.30}$$

or considering equation (2.81) this expression becomes

$$\mathbf{R}^T\,\mathbf{F}\,\mathbf{R}\,\mathbf{I} = \mathbf{0} \tag{3.31}$$

as $\mathbf{R}$ is known and non-singular, then

$$\mathbf{F}\,\mathbf{R}\,\mathbf{I} = \mathbf{0} \tag{3.32}$$

and a simple relationship has been obtained giving the values of the diagonal coefficients f_{ii} in terms of the off-diagonal terms f_{ij}.

3.2.2 Matrix G for Constant Elements

For the particular case of constant element, an entry of the matrix $\mathbf{G}$ is given by the following integral:

$$g_{ij} = \lim_{\varepsilon \to 0} \int_{\Gamma_j'} u^{*i} \; d\Gamma \tag{3.33}$$

where both i and j vary from 1 to N.

As mentioned before, in the classical direct boundary element formulation there is a matrix usually called $\mathbf{G}$, which is defined exactly in the same manner as the matrix $\mathbf{G}$ in this hybrid formulation. Therefore procedures similar to those adopted in the conventional boundary element method [40, 64] will be followed here.

By considering expression (2.49) which gives the fundamental solution, a coefficient of the matrix $\mathbf{G}$ can be written as

$$g_{ij} = \lim_{\varepsilon \to 0} \frac{1}{2\pi} \int_{\Gamma_j'} \ln \left(\frac{1}{r^i} \right) \; d\Gamma \tag{3.34}$$

According to this expression the off-diagonal terms are regular integrals and can be obtained through the evaluation of the following integral:

$$g_{ij} = \frac{1}{2\pi} \int_{\Gamma_j} \ln \left(\frac{1}{r^i} \right) \; d\Gamma \qquad i \neq j \tag{3.35}$$

which can be computed in this case either analytically or numerically. A Gauss quadrature rule can be used to evaluate them numerically, for instance.

The main diagonal terms, obtained when integrating over the element i, which includes the source point i, are given by

$$g_{ii} = \lim_{\varepsilon \to 0} \frac{1}{2\pi} \int_{\Gamma_i'} \ln(\frac{1}{r^i}) \; d\Gamma \tag{3.36}$$

In the limit they involve a logarithmic type of singularity, which is weak, therefore it is also equal to

$$g_{ii} = \frac{1}{2\pi} \int_{\Gamma_i} \ln\left(\frac{1}{r^i}\right) \, d\Gamma \tag{3.37}$$

and hence either an analytical or an appropriate numerical integration can be applied in its computation.

In the case of constant elements the analytical integration is very simple and will be preferred herein.

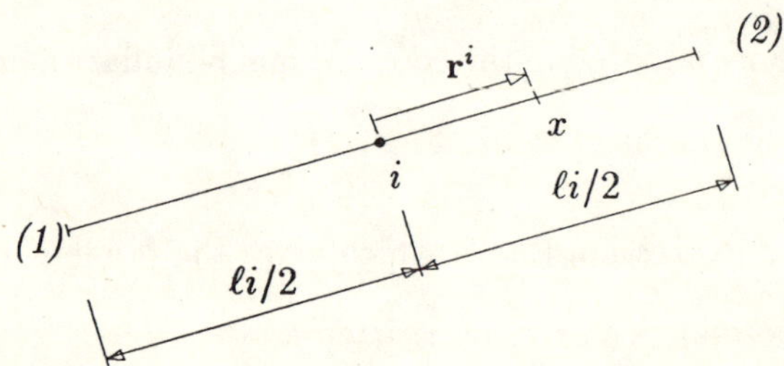

Figure 3.8: The element includes the source point i

According to figure 3.8 and taking symmetry into account, g_{ii} can be written as

$$
\begin{aligned}
g_{ii} &= \frac{1}{2\pi} \int_{(1)}^{(2)} \ln\left(\frac{1}{r^i}\right) \, d\eta \\
&= \frac{1}{\pi} \int_{(node\ i)}^{(2)} \ln\left(\frac{1}{r^i}\right) \, d\eta
\end{aligned} \tag{3.38}
$$

Now the homogeneous coordinate η defined over the element in figure 3.2 will be used to perform the integration. Therefore it is possible to write

$$r^i = \frac{\ell_i}{2} \, |\eta| \tag{3.39}$$

$$d\Gamma = \frac{\ell_i}{2} \, d\eta \tag{3.40}$$

and by substituting these two expressions into equation (3.38) g_{ii} becomes

$$g_{ii} = \frac{\ell_i}{2\pi} \int_0^1 \ln\left(\frac{2}{\eta\,\ell_i}\right) d\eta$$

$$= \frac{\ell_i}{2\pi}\left[\int_0^1 \ln\left(\frac{2}{\ell_i}\right) + \int_0^1 \ln\left(\frac{1}{\eta}\right) d\eta\right] \tag{3.41}$$

Now it will be shown that the last integral in the previous expression is equal to 1.

$$\int_0^1 \ln\left(\frac{1}{\eta}\right) d\eta = -\lim_{\varepsilon\to 0} \int_\varepsilon^1 \ln(\eta)d\eta$$

$$= 1 + \lim_{\varepsilon\to 0}\left(\varepsilon\ln\varepsilon - \varepsilon\right) \tag{3.42}$$

and since

$$\lim_{\varepsilon\to 0}\left(\varepsilon\ln\varepsilon\right) = 0$$

the value of the integral is given by

$$\int_0^1 \ln\left(\frac{1}{\eta}\right) d\eta = 1 \tag{3.43}$$

The main diagonal terms of matrix $\mathbf{G}$ can then be evaluated as

$$g_{ii} = \frac{\ell_i}{2\pi}\left[\ln\left(\frac{2}{\ell_i}\right) + 1\right] \tag{3.44}$$

3.2.3 Matrix L for Constant Elements

When dealing with constant elements the matrix $\mathbf{L}$ defined in equation (2.69) is a diagonal matrix and does not require any integration to be carried out. It is simply given by

$$\mathbf{L} = \begin{bmatrix} \ell_1 & & & \\ & \ell_2 & & \\ & & \ddots & \\ & & & \ell_N \end{bmatrix} \tag{3.45}$$

where ℓ_k $(k = 1, 2, \ldots, N)$ are the element lengths. Consequently it is not necessary to store this matrix and the related matrix operations become simply scalar operations.

3.2.4 Equivalent Nodal Fluxes

The vector of equivalent nodal fluxes $\mathbf{Q}$ is defined by equation (2.83), i. e.

$$\mathbf{Q} = \overline{\mathbf{Q}} + \mathbf{R}^T \mathbf{B} \qquad (3.46)$$

For constant elements $\overline{\mathbf{Q}}$ is directly obtained as a consequence of the assumption made of constant flux over an element. The prescribed equivalent flux — $\overline{Q}_i$ — at node i which is the midpoint of element Γ_i is given by

$$\overline{Q}_i = \overline{q}_i \, \ell_i \qquad (3.47)$$

in which $\overline{q}_i$ is the average prescribed flux on Γ_i; the summation convention is not implied here.

For Laplace's equation, i. e. in the absence of internal sources, vector $\mathbf{B}$ is the null vector. When the problem has internal sources, which corresponds to the case of Poisson's equation, the vector $\mathbf{B}$ has to be computed. Its evaluation is treated in section 3.4.

3.3 The Quadratic Element

This section is concerned with the derivation of all the matrices needed for the computer implementation of the hybrid boundary element formulation presented in chapter 2, when a quadratic isoparametric element is used to discretize the boundary (figure 3.9).

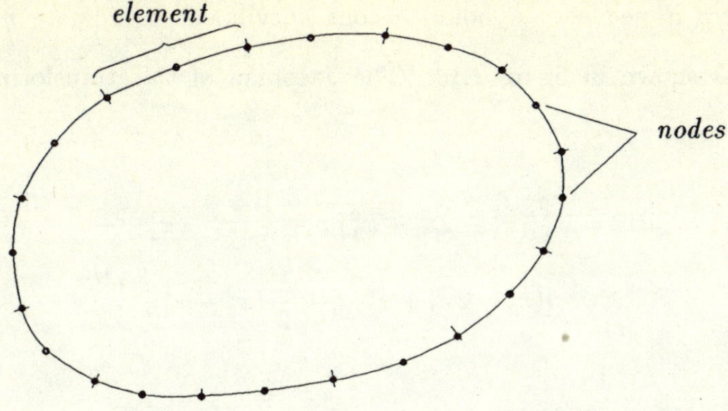

Figure 3.9: Boundary approximated by quadratic elements

This type of element is useful when dealing with curved boundaries because the geometry can thus be better represented. The use of quadratic elements is also important when the actual solution can not be well represented along the element through lower order functions. Furthermore in the case of lower order functions, the convergence of the solution can be very slow and sometimes cannot be obtained. The use of quadratic or higher order elements usually leads to convergence.

Three nodes per element are needed in order to define its geometry and the functions $\tilde{u}$ and $\tilde{q}$, which represent the approximate potential and flux on the element. These nodes are chosen to be the two extreme points of the element — node (1) and node (3) — and its midpoint, node (2) (figure 3.10). Therefore a C^0 continuity is obtained for the functions defining the geometry and $\tilde{u}$ and $\tilde{q}$.

The application of quadratic elements requires integrations along the boundary and to this end a transformation from Cartesian to curvilinear coordinates is needed. A homogeneous curvilinear coordinate η is then defined as shown in figure 3.10. The Jacobian of this transformation is given by

$$|J| = \left\{ [(x_1^1 - 2x_1^3 + x_1^2)\,\eta + \frac{1}{2}\,(x_1^2 - x_1^1)]^2 + \right.$$
$$\left. [(x_2^1 - 2x_2^3 + x_2^2)\,\eta + \frac{1}{2}\,(x_2^2 - x_2^1)]^2 \right\}^{1/2} \tag{3.48}$$

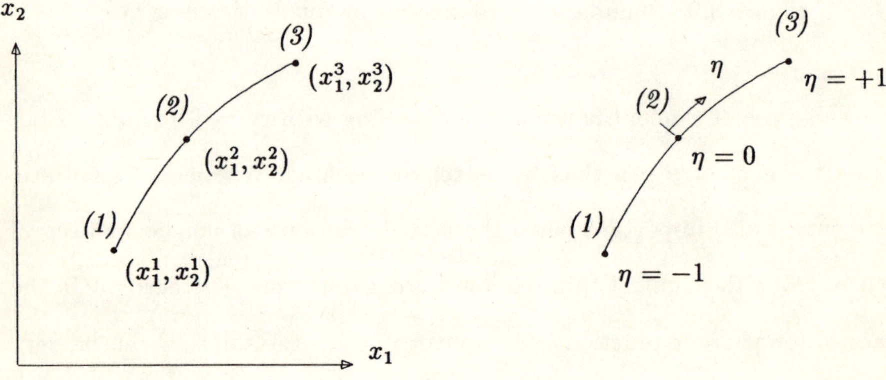

Figure 3.10: Transformation of coordinates

The independent variables on the boundary — $\tilde{u}$ and $\tilde{q}$ — are given in terms of quadratic Lagrange polynomials as interpolation functions and nodal values of this variables (see figure 3.11). Thus, along an element, they are written as

$$\tilde{u} = \varphi_1 u^1 + \varphi_2 u^2 + \varphi_3 u^3 = [\varphi_1 \ \varphi_2 \ \varphi_3] \left\{ \begin{array}{c} u^1 \\ u^2 \\ u^3 \end{array} \right\} \tag{3.49}$$

$$\tilde{q} \;=\; \varphi_1 q^1 + \varphi_2 q^2 + \varphi_3 q^3 = [\varphi_1 \;\; \varphi_2 \;\; \varphi_3] \left\{ \begin{array}{c} q^1 \\ q^2 \\ q^3 \end{array} \right\} \qquad (3.50)$$

where u^i and q^i $(i = 1, 2, 3)$ are, respectively, the values of $\tilde{u}$ and $\tilde{q}$ at the node i of the element. *Notice that although these are nodal values for the boundary variables $\tilde{u}$ and $\tilde{q}$, the tilde has been dropped in the notation for simplicity.*

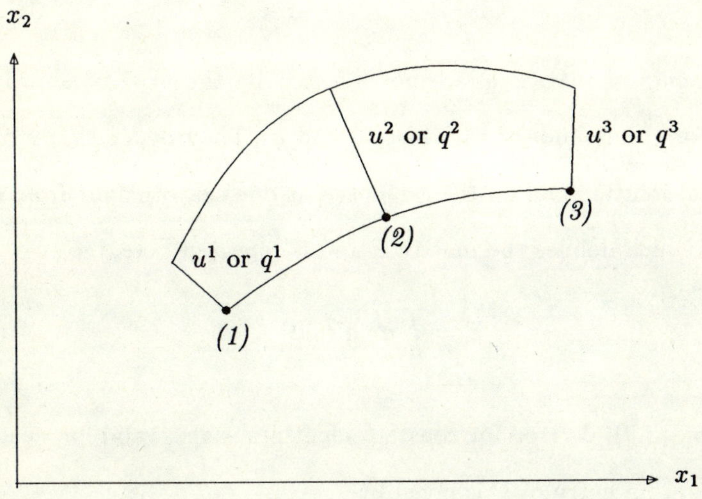

Figure 3.11: Approximation for $\tilde{u}$ and $\tilde{q}$

The shape functions φ_1, φ_2 and φ_3 are defined in terms of the homogeneous coordinate η as follows:

$$\varphi_1 = \frac{1}{2}\,\eta\,(\eta - 1) \qquad \varphi_2 = (1 - \eta)\,(1 + \eta) \qquad \varphi_3 = \frac{1}{2}\,\eta\,(\eta + 1) \qquad (3.51)$$

As the element is isoparametric its geometry is also approximated through the same shape functions. Therefore the Cartesian coordinates of a point on the element — x_1 and x_2 — are given in terms of the homogeneous coordinate η as shown below:

$$x_1 \;=\; \varphi_1 x_1^1 + \varphi_2 x_1^2 + \varphi_3 x_1^3 \tag{3.52}$$

$$x_2 \;=\; \varphi_1 x_2^1 + \varphi_2 x_2^2 + \varphi_3 x_2^3 \tag{3.53}$$

where x_1^i and x_2^i ($i = 1, 2, 3$) are the coordinates of the node i referred to the Cartesian global system (figure 3.10)

3.3.1   Matrix F for Quadratic Elements

The coefficients of matrix $\mathbf{F}$ are not affected by the type of approximation used for the boundary variables, $\tilde{u}$ and $\tilde{q}$. They depend only on the fundamental solution and on the geometry, as one can conclude from equation (2.67) which defines the matrix $\mathbf{F}$ and is repeated here, i. e.

$$\mathbf{F} = \int_{\Gamma} \mathbf{u}^* \, \mathbf{q}^{*T} \, d\Gamma \tag{3.54}$$

Equation (3.9), derived for constant elements is also valid for quadratic elements, i. e. an entry of $\mathbf{F}$ is given by

$$f_{ij} = -\,\frac{\rho}{4\pi^2} \lim_{\varepsilon \to 0} \sum_{k=1}^{E} \int_{\Gamma'_k} \ln\left(\frac{1}{r^i}\right) \frac{1}{r^j} \frac{\partial r^j}{\partial n} \, d\Gamma(x) \tag{3.55}$$

where E is the number of boundary elements. According to the last expression, an integration along the whole boundary has to be performed to compute a coefficient f_{ij}. To this end five different cases should be considered, depending on the position of the sources i and j in relation to the element.

Case 1: *Integration over an element which includes neither source point i nor source point j.*

This case involves the non-singular integral given by

$$\int_{\Gamma_k} \ln\left(\frac{1}{r^i}\right) \frac{1}{r^j} \frac{\partial r^j}{\partial n} \, d\Gamma(x) \qquad (3.56)$$

where the variables involved are shown in figure 3.12.

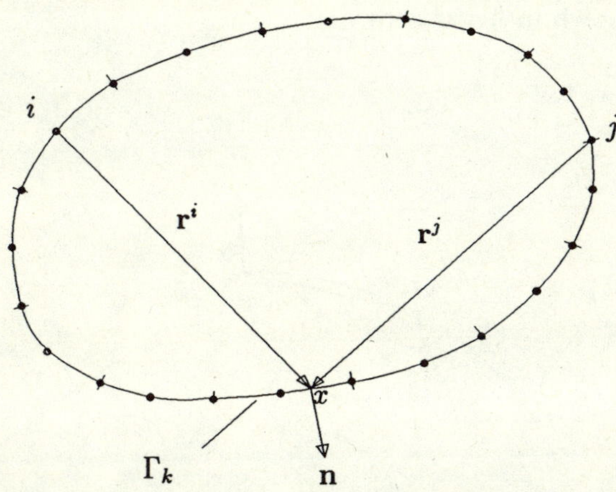

Figure 3.12: Sources i and j outside the element

These integrals can be evaluated numerically by using a standard Gauss quadrature rule. Care should be taken when any of the sources i and j or both of them are close to the element over which the integration is being performed. The rapid variation of the integrand when a singularity is approached can affect the results After a series of tests, it was found that 12 integration points give good accuracy, when the distance between one source point and the nearest point on the element is greater than the distance from the midnode to the extreme node closer to it.

Case 2: *Integration over an element which includes only the source point i.*

The integral to be performed in this situation is herein denoted by $\Im^i$ and is expressed by the following formula:

$$\Im^i = \lim_{\varepsilon \to 0} \int_{\Gamma'_k} \ln\left(\frac{1}{r^i}\right) \frac{1}{r^j} \frac{\partial r^j}{\partial n} \, d\Gamma(x) \qquad (3.57)$$

with the boundary Γ'_k $(\Gamma'_k \to \Gamma_k$ when $\varepsilon \to 0)$ and the vectors $\mathbf{r}^i$ and $\mathbf{r}^j$ shown in figure 3.13.

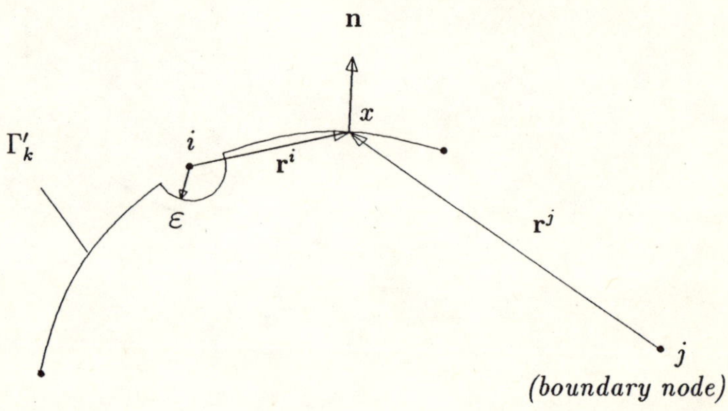

Figure 3.13: Singularity i on the element Γ_k

In the limit, when $\varepsilon = 0$, this integral contains a logarithmic type of singularity which is weak and hence the integral is well defined. It is then possible to write

$$\Im^i = \int_{\Gamma_k} \ln\left(\frac{1}{r^i}\right) \frac{1}{r^j} \frac{\partial r^j}{\partial n} \, d\Gamma(x) \qquad (3.58)$$

A suitable numerical scheme can then be used to obtain the value for this integral, as mentioned previously when dealing with constant elements (section 3.2 case 2).

When the source point i is located at an end node, i. e. node (1) or node (2) (figure 3.10), similar integration have to be performed for the adjacent element which also includes the source i and the results should be added to give the final value of $\Im^i$. Obviously when the source point i is at the central node the integral is evaluated along the element under consideration only.

Case 3: *Integration over an element which includes only source point j.*

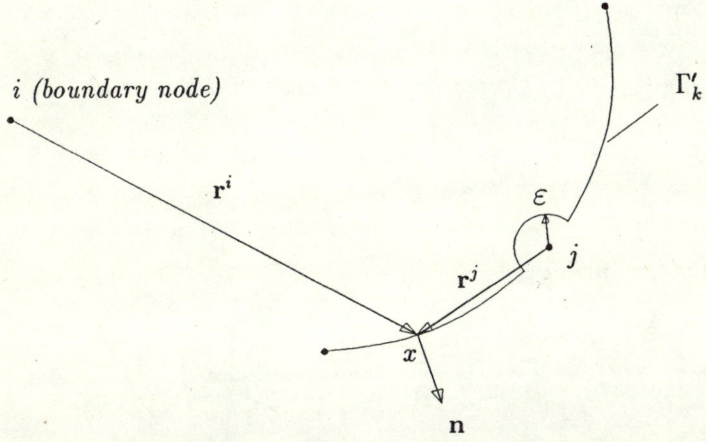

Figure 3.14: Singularity j on the element Γ_k

Herein the following integral should be computed:

$$\Im^j = \lim_{\varepsilon \to 0} \int_{\Gamma'_k} \ln\left(\frac{1}{r^i}\right) \frac{1}{r^j} \frac{\partial r^j}{\partial n} \, d\Gamma(x) \qquad (3.59)$$

where the element Γ_k $(\Gamma'_k \to \Gamma_k$ when $\varepsilon \to 0)$ includes the source j and the other variables in the integrand are shown in figure 3.14.

Along the boundary this integral is not singular, however it presents a discontinuity when approaching it. Consequently $\Im^j$ is computed by adding the results for the discontinuity or jump and the value of the integral over the element, i. e.

$$\Im^j = \int_{\Gamma_k} \ln\left(\frac{1}{r^i}\right) \frac{1}{r^j} \frac{\partial r^j}{\partial n} \, d\Gamma(x) + \lim_{\varepsilon \to 0} \int_{\Gamma_\varepsilon} \ln\left(\frac{1}{r^i}\right) \frac{1}{r^j} \frac{\partial r^j}{\partial n} \, d\Gamma(x) \quad (3.60)$$

where Γ_ε is the arc of radius ε, which is shown in figure 3.14.

The first integral on the right-hand side is well defined and can be computed by following the same procedures used in case 1. The second term corresponds to the discontinuity and will be obtained analytically.

Two possibilities should be considered in the evaluation of the jump, as follows:

i) *Source point j located at an end node of the element.*

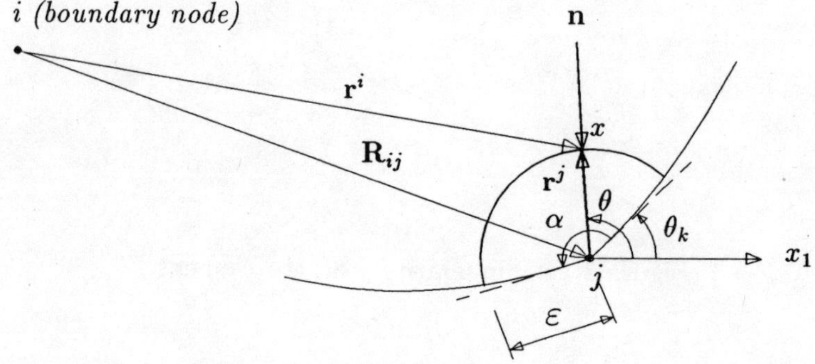

Figure 3.15: Singularity at an end node of the element.

According to figure 3.15, it is possible to write the following relationships:

$$r^i = |\mathbf{R}_{ij} + \mathbf{r}^j| \tag{3.61}$$

$$r^j = \varepsilon \tag{3.62}$$

$$\frac{\partial r^j}{\partial n} = -1 \tag{3.63}$$

$$d\Gamma = -\varepsilon \, d\theta \tag{3.64}$$

By introducing the last equations into the expression for the jump, i. e. the second integral on the right-hand side of equation (3.60), this becomes

$$\lim_{\varepsilon \to 0} \int_{\Gamma_\varepsilon} \ln\left(\frac{1}{r^i}\right) \frac{1}{r^j} \frac{\partial r^j}{\partial n}\, d\Gamma = \lim_{\varepsilon \to 0} \int_{\theta_k+\alpha}^{\theta_k} \ln\left(\frac{1}{|\mathbf{R}_{ij} + \mathbf{r}^j|}\right) \frac{1}{\varepsilon} \cdot \varepsilon\, d\theta$$

$$= -\alpha \ln\left(\frac{1}{R_{ij}}\right) \qquad (3.65)$$

where θ_k is the angle at node j between the x_1 axis and the element after node j and α is the angle at j between the two elements, as shown in figure 3.15.

ii) *Source point j located at the midpoint of the element.*

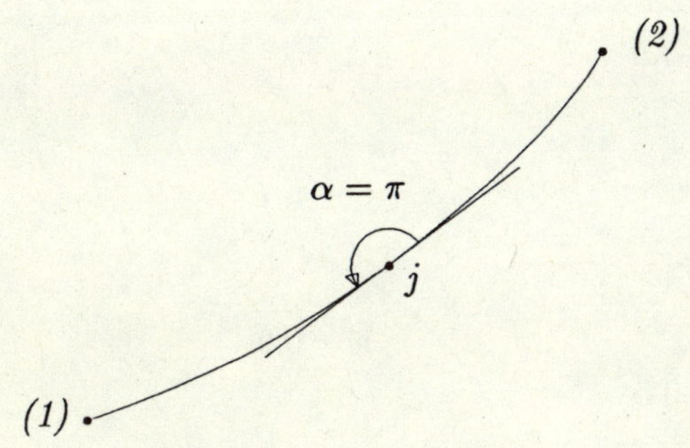

Figure 3.16: Singularity j at the midnode of the element

The jump in this situation can be computed simply by replacing the value of α by π in equation (3.65) as one can conclude by comparing the two cases, **i** and **ii**. See figures 3.15 and 3.16

Case 4: *Integration over an element which includes both non-coincident source points i and j.*

This is the most delicate situation because two different types of singularity occur for the same element. The integrals to be obtained are still given by equation (3.9), i.e.

$$f_{ij} = -\frac{\rho}{4\pi^2} \lim_{\epsilon \to 0} \sum_{k=1}^{n} \int_{\Gamma'_k} \ln\left(\frac{1}{r^i}\right) \frac{1}{r^j} \frac{\partial r^j}{\partial n} \, d\Gamma(x) \qquad (3.66)$$

The element is subdivided in such a way that only one singularity is included in each integral, as shown in figure 3.17.

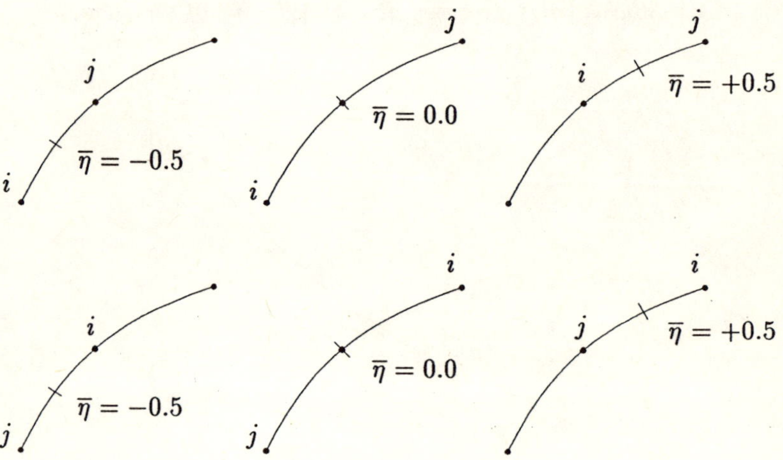

Figure 3.17: Both singularities i and j occurring on the same element

The proximity of a singularity can affect the results of the singular integrals. Therefore the point where the element should be subdivided, namely the value of $\overline{\eta}$ shown in figure 3.17, has to be chosen carefully and enough integration points should be used to guarantee the accuracy and convergence of the results.

According to figure 3.17 there are six different situations to be considered depending on the position of the sources i and j.

When the integration is performed over a boundary which includes the source i, a logarithmic singularity occurs and a special scheme to deal with a logarithmic type of singularity should be used [67], [68]. This situation is similar to case 2.

The situation in which the source j is included in the region where the integral is defined is similar to case 3. Consequently the integral on the boundary is not singular and can be computed through a normal Gaussian quadrature rule, provided the discontinuity in the integral when approaching the boundary has been properly evaluated. Herein it is equal to the jump studied in case 3.

Case 5: *Integration over an element including both source points i and j, which are coincident.*

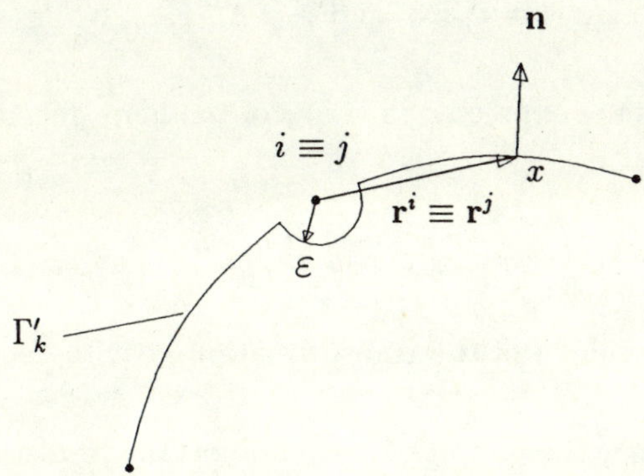

Figure 3.18: Singularities i and j at the same node on the element

This case corresponds to the evaluation of the main diagonal of matrix $\mathbf{F}$ whose coefficients are given by

$$f_{ii} = \lim_{\varepsilon \to 0} \int_{\Gamma'_k} \ln \left(\frac{1}{r^i}\right) \frac{1}{r^i} \frac{\partial r^i}{\partial n} \, d\Gamma \qquad (3.67)$$

Exactly the same physical consideration used for constant elements (section 3.2) can be adopted here to evaluate these terms and avoid their direct computation.

3.3.2 Matrix G for Quadratic Elements

Matrix $\mathbf{G}$ is a N by N square matrix defined through equation (2.68). Consequently for the isoparametric quadratic element being developed in this section, when the nodes and elements are numbered sequentially, its coefficients are given as shown in the next page.

Thus it is necessary to compute integrals of the type

$$\int_{\Gamma_k} \varphi_l \, u^{*i} \, d\Gamma \qquad (3.68)$$

where $k = 1, 2, \ldots, E$, $\quad l = 1, 2, 3$ and $i = 1, 2, \ldots, N$.

These integrals are expressed in terms of the homogeneous coordinate η as

$$\int_{\Gamma_k} \varphi_l \, u^{*i} \, d\Gamma = \int_{(1)}^{(2)} \varphi_l \, u^{*i} \, |J| \, d\eta \qquad (3.69)$$

where $|J|$ is the Jacobian of the transformation given by equation (3.48).

When computing these integrals it is important to distinguish between the singular and non-singular integrals.

Matrix G for Quadratic Elements

$$
\begin{bmatrix}
\int_{\Gamma_1}\varphi_1 u_1^* d\Gamma + \int_{\Gamma_{Ne}}\varphi_3 u_1^* d\Gamma & \int_{\Gamma_1}\varphi_2 u_1^* d\Gamma & \int_{\Gamma_1}\varphi_3 u_1^* d\Gamma + \int_{\Gamma_2}\varphi_1 u_1^* d\Gamma & \int_{\Gamma_2}\varphi_2 u_1^* d\Gamma & \cdots & \int_{\Gamma_{Ne-1}}\varphi_3 u_1^* d\Gamma + \int_{\Gamma_{Ne}}\varphi_1 u_1^* d\Gamma & \int_{\Gamma_{Ne}}\varphi_2 u_1^* d\Gamma \\[2ex]
\int_{\Gamma_1}\varphi_1 u_2^* d\Gamma + \int_{\Gamma_{Ne}}\varphi_3 u_2^* d\Gamma & \int_{\Gamma_1}\varphi_2 u_2^* d\Gamma & \int_{\Gamma_1}\varphi_3 u_2^* d\Gamma + \int_{\Gamma_2}\varphi_1 u_2^* d\Gamma & \int_{\Gamma_2}\varphi_2 u_2^* d\Gamma & \cdots & \int_{\Gamma_{Ne-1}}\varphi_3 u_2^* d\Gamma + \int_{\Gamma_{Ne}}\varphi_1 u_2^* d\Gamma & \int_{\Gamma_{Ne}}\varphi_2 u_2^* d\Gamma \\[2ex]
\vdots & \vdots & \vdots & \vdots & \cdots & \vdots & \vdots \\[2ex]
\int_{\Gamma_1}\varphi_1 u_n^* d\Gamma + \int_{\Gamma_{Ne}}\varphi_3 u_n^* d\Gamma & \int_{\Gamma_1}\varphi_2 u_n^* d\Gamma & \int_{\Gamma_1}\varphi_3 u_n^* d\Gamma + \int_{\Gamma_2}\varphi_1 u_n^* d\Gamma & \int_{\Gamma_2}\varphi_2 u_n^* d\Gamma & \cdots & \int_{\Gamma_{Ne-1}}\varphi_3 u_n^* d\Gamma + \int_{\Gamma_{Ne}}\varphi_1 u_n^* d\Gamma & \int_{\Gamma_{Ne}}\varphi_2 u_n^* d\Gamma
\end{bmatrix}
$$

i) *Non-singular integrals.*

This situation occurs when the source point i is not located on the element Γ_k, where the integration is being evaluated. A normal Gaussian quadrature rule can then be applied. The influence of the proximity of a singularity should be considered when selecting the number of integration points to be adopted.

ii) *Singular integrals.*

In this case, the element Γ_k includes the source point i. As the functions u^{*i} are fundamental solutions given by equation (3.5), the integral contains a logarithmic type of singularity. Therefore an appropriate numerical scheme should be used.

3.3.3   Matrix L for Quadratic Elements

According to expression (2.69) that defines the matrix $\mathbf{L}$ and taking into account the type of approximation used for $\tilde{u}$ and $\tilde{q}$, equations (3.49) and (3.50), it is possible to represent the coefficients of matrix $\mathbf{L}$ as shown in the next page. This matrix does not present any singularity and its numerical evaluation is a simple task.

3.3.4   Equivalent Nodal Fluxes

Equation (2.83) defines the vector of equivalent nodal fluxes, i. e.

$$Q = \overline{Q} + \mathbf{R}^T \, \mathbf{B} \qquad (3.70)$$

The vector $\mathbf{B}$ for quadratic elements is similar to vector $\mathbf{B}$ for constant elements and will be discussed in section 3.4.

Matrix L for Quadratic Elements

$$
\begin{bmatrix}
\int_{\Gamma_1} \varphi_1^2 d\Gamma + \int_{\Gamma_{Ne}} \varphi_3^2 d\Gamma & \int_{\Gamma_1} \varphi_1\varphi_2 d\Gamma & \int_{\Gamma_1} \varphi_1\varphi_3 d\Gamma & 0 & 0 & \cdots & \int_{\Gamma_{Ne}} \varphi_1\varphi_3 d\Gamma & \int_{\Gamma_{Ne}} \varphi_2\varphi_3 d\Gamma \\[2ex]
\int_{\Gamma_1} \varphi_1\varphi_2 d\Gamma & \int_{\Gamma_1} \varphi_2^2 d\Gamma & \int_{\Gamma_1} \varphi_2\varphi_3 d\Gamma & 0 & 0 & \cdots & 0 & 0 \\[2ex]
\int_{\Gamma_1} \varphi_1\varphi_3 d\Gamma & \int_{\Gamma_1} \varphi_2\varphi_3 d\Gamma & \int_{\Gamma_1} \varphi_3^2 d\Gamma + \int_{\Gamma_2} \varphi_1^2 d\Gamma & \int_{\Gamma_2} \varphi_1\varphi_2 d\Gamma & \int_{\Gamma_2} \varphi_1\varphi_3 d\Gamma & \cdots & 0 & 0 \\[2ex]
0 & 0 & \int_{\Gamma_2} \varphi_1\varphi_2 d\Gamma & \int_{\Gamma_2} \varphi_2^2 d\Gamma & \int_{\Gamma_2} \varphi_2\varphi_3 d\Gamma & \cdots & 0 & 0 \\[2ex]
0 & 0 & \int_{\Gamma_2} \varphi_1\varphi_3 d\Gamma & \int_{\Gamma_2} \varphi_2\varphi_3 d\Gamma & \int_{\Gamma_2} \varphi_3^2 d\Gamma + \int_{\Gamma_3} \varphi_1^2 d\Gamma & \cdots & 0 & 0 \\[2ex]
\vdots & \vdots & \vdots & \vdots & \vdots & \vdots & \vdots & \vdots \\[2ex]
\int_{\Gamma_{Ne}} \varphi_2\varphi_3 d\Gamma & 0 & 0 & 0 & 0 & \cdots & \int_{\Gamma_{Ne}} \varphi_1\varphi_2 d\Gamma & \int_{\Gamma_{Ne}} \varphi_2^2 d\Gamma
\end{bmatrix}
$$

Suppose that the prescribed flux on the element Γ_e — $\bar{q}_e$ — can be represented through quadratic interpolation functions, in the same way the fluxes on the boundary, $\tilde{q}$, are approximated (3.50). It is then possible to write

$$\bar{q}_e = \varphi_1\,\bar{q}_e^1 + \varphi_2\,\bar{q}_e^2 + \varphi_3\,\bar{q}_e^3 = [\varphi_1 \ \varphi_2 \ \varphi_3] \left\{ \begin{array}{c} \bar{q}_e^1 \\ \bar{q}_e^2 \\ \bar{q}_e^3 \end{array} \right\} \tag{3.71}$$

where $\bar{q}_e^1$, $\bar{q}_e^2$ and $\bar{q}_e^3$ are, respectively, the values of the prescribed flux at nodes (1) , (2) and (3) on Γ_e.

Then by substituting equation (3.71) into expression (2.70) the contribution of the element Γ_e for the equivalent nodal fluxes can be expressed as

$$\left\{ \begin{array}{c} \bar{Q}_e^1 \\ \bar{Q}_e^2 \\ \bar{Q}_e^3 \end{array} \right\} = \int_{\Gamma_e} \left\{ \begin{array}{c} \varphi_1 \\ \varphi_2 \\ \varphi_3 \end{array} \right\} [\varphi_1 \ \varphi_2 \ \varphi_3] \, d\Gamma \left\{ \begin{array}{c} \bar{q}_e^1 \\ \bar{q}_e^2 \\ \bar{q}_e^3 \end{array} \right\} \tag{3.72}$$

where $\bar{Q}_e^i$ $(i = 1, 2, 3)$ represents the contribution of the element Γ_e to the equivalent prescribed flux at node i.

Notice that all these integrals have been already computed in the evaluation of matrix **L**.

3.4   The Vector B

In this section, the evaluation of vector **B** will be discussed. This vector has to be computed when a Poisson type equation is being solved. In engineering practice this happens, for instance, in heat conduction problems presenting internal heat generation or in the analysis of well pumping in confined aquifers.

The vector $\mathbf{B}$ is defined by equation (2.71) which involves a domain integral. For the type of fundamental solution adopted here, the integral is well defined and as such can be written as

$$\mathbf{B} = \int_\Omega b(x)\, \mathbf{u}^* \, d\Omega(x) \qquad (3.73)$$

where b was defined in section 2.3 and x is a point in the domain Ω.

Each component B_i of $\mathbf{B}$ corresponds to a particular position of the source point i on the boundary, i. e.

$$B_i = \int_\Omega b(x)\, u^{*i} \, d\Omega(x) \qquad (3.74)$$

where u^{*i} $(i = 1, 2, \ldots, N)$ is the fundamental solution at the point x in the domain, due to a unit source at the boundary node i.

The need for performing domain integrals is the main drawback in a boundary element formulation because of the necessity to discretize the domain and the consequent lost of the pure boundary feature in the method. The consequences of this fact is that the input data has to include the discretization of the domain and time consuming domain integrals have to be performed. That is why researchers involved with the conventional boundary element method have been actively trying to avoid such domain integrals. They either transform them into boundary integrals or, alternatively, by using a particular solution or by changing variables, the original equation is converted into an equivalent equation for which a fundamental solution is available. Some of these techniques depend intrinsically on the boundary element formulation. Others do not keep any relation to the kind of method applied and involve purely mathematical concepts and relationships.

Domain integrals similar to those in expression (3.74) appear in the classical boundary element method formulation for Poisson's equation. Therefore techniques for their evaluation without discretization of the domain, which are independent of the particular formulation adopted, can be used in this hybrid boundary element approach to obtain the components of vector **B**. Some of them will be briefly described in this section.

Notice that for this hybrid displacement boundary elements formulation the computation of potentials and fluxes in the domain does not require any kind of integration even in the case of Poisson's equation (see section 2.5.6).

In what follows some procedures that can be used to compute the vector **B** will be described.

Direct computation of the domain integrals

For direct computation of the domain integrals in equation (3.74), the domain should be subdivided into regions of integration called cells (figure 3.19). A numerical integration scheme such as Gaussian quadrature formulae can then be applied.

Supposing that the total number of cells needed to describe the whole domain is C and that Ω_l is a generic cell, it is possible to write a component B_i as

$$B_i = \sum_{l=1}^{C} \int_{\Omega_l} b \, u^{*i} \, d\Omega \qquad (3.75)$$

The numerical integration scheme can now be applied for every cell and the results added up.

Concentrated sources

The evaluation of **B** in this case is very simple. For one concentrated source – S_ζ – at a point ζ the function f (see equation (2.21)) is given by

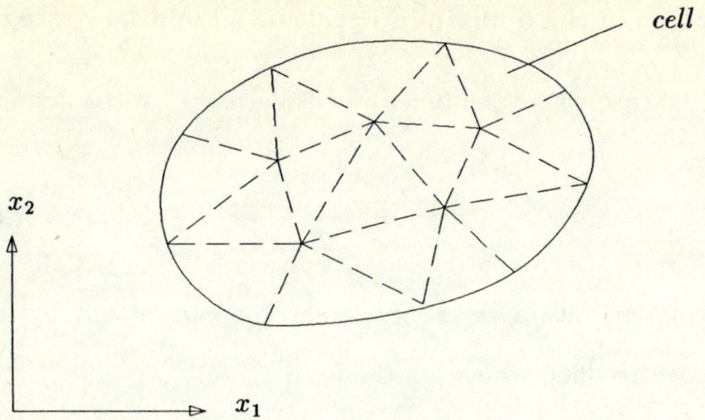

Figure 3.19: Domain discretization

$$f = S_\zeta \, \Delta(\zeta, x) \tag{3.76}$$

where S_ζ is the magnitude of the source at ζ and $\Delta(\zeta, x)$ is the Dirac delta function (see section 2.5.1). Therefore b can be written as

$$b = -\rho \, f = -\rho \, S_\zeta \, \Delta(\zeta, x) \tag{3.77}$$

and a component B_i becomes

$$B_i = \int_\Omega -\rho \, S_\zeta \, \Delta(\zeta, x) \, u^{*i}(x) \, d\Omega(x) \tag{3.78}$$

By applying Dirac delta function properties it can be recast as

$$B_i = -\rho \, S_\zeta \, u^{*i}(\zeta) \tag{3.79}$$

For the case of several concentrated sources the superposition of effects can be applied.

Transformation of the domain integrals into boundary integrals

Herein only the case where the function b is harmonic in the domain Ω will be presented, i. e.

$$\nabla^2 b = 0 \tag{3.80}$$

The transformation into a boundary integral is carried out by using Green's second identity [66], which has the form

$$\int_\Omega \left(b \, \nabla^2 v^* - v^* \, \nabla^2 b \right) \, d\Omega = \int_\Gamma \left(b \, \frac{\partial v^*}{\partial n} - v^* \, \frac{\partial b}{\partial n} \right) \, d\Gamma \tag{3.81}$$

This expression can be applied to take the volume integral in (3.74) to the boundary if there is a function v^* such that

$$\nabla^2 v^* = u^* \tag{3.82}$$

This function v^* is the fundamental solution of the bi-harmonic equation as shown bellow

$$\nabla^2 u^* = -\Delta(\xi, x) = \nabla^2(\nabla^2 v^*) = \nabla^4 v^* = -\Delta(\xi, x) \tag{3.83}$$

This solution is a well known fundamental solution used in plate bending which is given for two-dimensional domains by

$$v^* = \frac{r^2}{8 \, \pi} \left[\ln \left(\frac{1}{r} \right) + 1 \right] \tag{3.84}$$

where r is the distance between the source point ξ and the field point x.

Substituting equations (3.80) and (3.82) into equation (3.81), gives

$$\int_\Omega b \, u^* \, d\Omega = \int_\Gamma (b \, \frac{\partial v^*}{\partial n} - v^* \, \frac{\partial b}{\partial n}) \, d\Gamma \tag{3.85}$$

And finally by using this last equation, the component B_i of vector $\mathbf{B}$ can be expressed in terms of two boundary integrals as follows:

$$\int_\Omega b \, u^{*i} \, d\Omega = \int_\Gamma (b \, \frac{\partial v_i^*}{\partial n} - v_i^* \, \frac{\partial b}{\partial n}) \, d\Gamma \tag{3.86}$$

where the subscript i means that the source point is located at the boundary node i.

Chapter 4

Elastostatics

4.1 Introduction

In this chapter, the hybrid displacement boundary element formulation for the solution of linear elastostatic problems will be presented. The existing similarities between the potential theory (also called scalar potential theory) and the theory of elasticity (also called vector potential theory [61]) results in the same format for this chapter as for chapter 2.

The main characteristics of the approach, discussed in chapter 2 for potential problems, remain the same for the case of elasticity. Thus the formulation is based on a generalized variational principle of the type used by Tong [1] to generate his assumed displacement hybrid model in finite elements. In that functional, the volume integral which represents the strain energy is integrated by parts and an ad hoc boundary hybrid functional is then generated. This functional forms the basis of the proposed approach.

There are three independent field variables involved: they are the displacement and tractions on the boundary and the displacement inside the domain.

Through a convenient approximation for the displacement in the domain it is possible to generate a stiffness-like formulation whose stiffness matrix is evaluated by integrations along the boundaries only. This can be achieved by using the so-called fundamental solution for elasticity to represent approximately that domain variable.

The variables defined only on the boundary, which are displacement and traction, are approximated in terms of the usual interpolation functions that can vary in accordance to the type of problem to be solved.

After introducing the three approximate field variables into the expression for the boundary hybrid functional and making its first variation equal to zero a set of three matrix equations is generated. The final system of algebraic equations is obtained by eliminating the variables other than boundary displacements from those equations. The resulting system of algebraic equations is therefore written in terms of a symmetric stiffness matrix and nodal values of displacements as unknowns, both defined on the boundary only.

4.2 Basic Relations in Linear Elastostatics

Let us consider a domain Ω, bounded by a closed surface Γ, which consists of a homogeneous, linear, elastic and isotropic material (figure 4.1).

The elastostatics problems defined in such domain can be represented by the following equations:

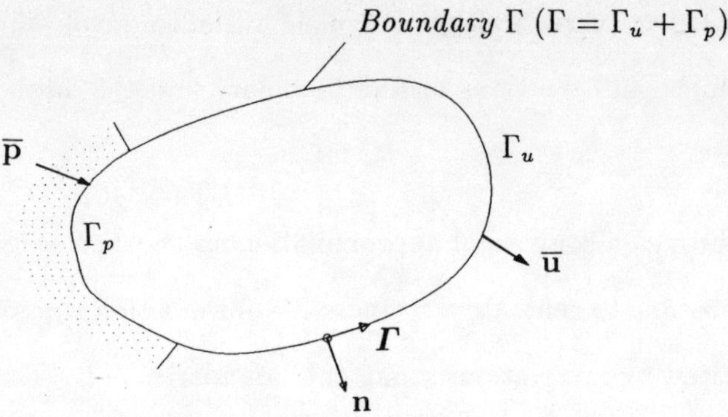

Figure 4.1: Definition of the domain Ω and boundary Γ

- *Strain-displacement equation*

If only small deformations are supposed to occur, the strain tensor ϵ_{ij} can be written, in terms of the displacement vector u_i, by

$$\epsilon_{ij} = \frac{1}{2}(u_{i,j} + u_{j,i}) \qquad (4.1)$$

- *Equilibrium conditions*

When the force equilibrium condition is fulfilled the stress tensor σ_{ij} satisfies the following equations:

$$\sigma_{ij,i} + b_j = 0 \qquad (4.2)$$

where the body forces b_j are external forces which act inside the body.

In the absence of distributed body or surface couples, the moment equilibrium condition requires the stress tensor to be symmetric, i. e.

$$\sigma_{ij} = \sigma_{ji} \qquad (4.3)$$

The vector of surface forces per unit area, or traction vector p_i, can be computed from the stress components σ_{ij} by the relation

$$p_i = \sigma_{ji}\, n_j \tag{4.4}$$

where n_j are the direction cosines of the outward normal $\mathbf{n}$ to the surface, at the point where p_i acts (figure 4.1).

- *Constitutive equations*

When there is no change in temperature Hooke's law, formulated in terms of shear modulus G and Poisson's ratio ν, relates the stress tensor components σ_{ij} and the strain components ϵ_{ij} leading to

$$\sigma_{ij} = C_{ijkl}\, \epsilon_{kl} \tag{4.5}$$

in which C_{ijkl} is the fourth-order isotropic tensor of elastic constants given by

$$C_{ijkl} = \frac{2\,G\,\nu}{1-2\nu}\, \delta_{ij}\, \delta_{kl} + G\,(\delta_{ik}\, \delta_{jl} + \delta_{il}\, \delta_{jk}) \tag{4.6}$$

- *Boundary conditions*

The boundary conditions for displacements and tractions are as follows:

$$\text{essential boundary conditions:} \quad u_i = \overline{u}_i \quad on \ \ \Gamma_u \tag{4.7}$$

$$\text{natural boundary conditions:} \quad p_i = \overline{p}_i \quad on \ \ \Gamma_p \tag{4.8}$$

Observe that the total bounding surface of the body Γ is equal to $\Gamma_u + \Gamma_p$ and that the bar indicates that those values are prescribed.

The equilibrium conditions can also be expressed in terms of displacements. The resulting equations are the well-known Navier equations which are very convenient when the boundary conditions are given is terms of displacements. These equations are obtained by substituting the strain-displacement equation (4.1) into (4.5) and taking the result into the equilibrium condition, given by (4.2). Thus Navier equations are written as

$$G\,u_{j,kk} + \frac{G}{1-2\nu}\,u_{k,kj} + b_j = 0 \tag{4.9}$$

In all the previous equations and throughout this chapter the indicial notation as defined in section ?? is applied.

4.3 Modified Variational Principle

The principle of minimum potential energy is a classical principle in mechanics of solids. It has been extensively applied in elasticity since the introduction of variational principles for the solution of structural mechanics problems.

The generalized variational principle, basis of this proposed hybrid formulation, will be derived in this section. It is obtained from the principle of minimum potential energy which can be stated as follows:

The solution of a problem in the small displacement theory of elasticity is the vector function u_i which minimizes the total potential energy Π given by

$$\Pi(u_i) = \int_\Omega (\tfrac{1}{2}\,\sigma_{ij}\,\epsilon_{ij} - b_i\,u_i)\,d\Omega - \int_{\Gamma_p} \bar{p}_i\,u_i\,d\Gamma \tag{4.10}$$

with the boundary condition (4.7), i. e.,

$$u_i = \bar{u}_i \qquad \text{on } \Gamma_u \tag{4.11}$$

which is repeated here for clarity.

In the expression for the functional Π (4.10) both tensors, σ_{ij} and ϵ_{ij}, are functions of the displacement vector u_i.

The present formulation has instead two independent displacement field variables. The first one is defined in the domain and called u_i; the second displacement field is denoted by $\tilde{u}_i$ and is only defined on the boundary. These two independent field variables will be introduced in the expression for the total potential energy to derive a more general principle. Consequently the compatibility condition between them needs to be introduced as a subsidiary condition. The variational form of the problem then becomes

$$\Pi_1(u_i) = \int_\Omega (\frac{1}{2}\,\sigma_{ij}\,\epsilon_{ij} - b_i\,u_i)\,d\Omega - \int_{\Gamma_p} \bar{p}_i\,\tilde{u}_i\,d\Gamma \qquad (4.12)$$

with the subsidiary compatibility condition,

$$u_i = \tilde{u}_i \qquad \text{on } \Gamma \qquad (4.13)$$

and the boundary condition

$$\tilde{u}_i = \overline{u}_i \qquad \text{on } \Gamma_u \qquad (4.14)$$

Now the subsidiary condition (4.13) will be introduced into the variational expression by means of a set of Lagrange multipliers λ_i. Thus the modified variational principle can now be written as:

The solution of the problem will be given by the satisfaction of the stationary condition of the multi-field functional Π_2 defined by the following expression:

$$\Pi_2(u_i, \tilde{u}_i, \lambda_i) = \int_\Omega (\frac{1}{2}\,\sigma_{ij}\,\epsilon_{ij} - b_i\,u_i)\,d\Omega + \int_\Gamma \lambda_i\,(\tilde{u}_i - u_i)\,d\Gamma - \int_{\Gamma_p} \bar{p}_i\,\tilde{u}_i\,d\Gamma$$

$$(4.15)$$

plus the boundary condition

$$\tilde{u}_i = \overline{u}_i \qquad \text{on } \Gamma_u \qquad (4.16)$$

Notice that the generalized functional now involves three independent field variables and the modified principle is a *stationary* principle and no longer a *minimum* principle.

The stationary conditions for the functional Π_2 are determined by setting to zero its first variation which can be obtained by taking variations with respect to the three independent variables u_i, $\tilde{u}_i$ and λ_i in equation (4.15) as follows:

$$\delta\Pi_2(u_i,\tilde{u}_i,\lambda_i) = \int_\Omega \left[\frac{1}{2}\left(\sigma_{ij}\,\delta\epsilon_{ij} + \epsilon_{ij}\,\delta\sigma_{ij}\right) - b_i\,\delta u_i\right] d\Omega + \int_\Gamma \lambda_i\,\delta\tilde{u}_i\,d\Gamma -$$
$$\int_\Gamma \lambda_i\,\delta u_i\,d\Gamma + \int_\Gamma (\tilde{u}_i - u_i)\,\delta\lambda_i\,d\Gamma - \int_{\Gamma_p} \bar{p}_i\,\delta\tilde{u}_i\,d\Gamma \qquad (4.17)$$

which after assuming the satisfaction of the boundary condition $\tilde{u}_i = \bar{u}_i$ on Γ_u, namely $\delta\tilde{u}_i = 0$ on Γ_u, becomes

$$\delta\Pi_2(u_i,\tilde{u}_i,\lambda_i) = \int_\Omega \left[\frac{1}{2}\left(\sigma_{ij}\,\delta\epsilon_{ij} + \epsilon_{ij}\,\delta\sigma_{ij}\right) - b_i\,\delta u_i\right] d\Omega - \int_\Gamma \lambda_i\,\delta u_i\,d\Gamma +$$
$$- \int_\Gamma (\tilde{u}_i - u_i)\,\delta\lambda_i\,d\Gamma + \int_{\Gamma_p} (\lambda_i - \bar{p}_i)\,\delta\tilde{u}_i\,d\Gamma \qquad (4.18)$$

From equation (4.5) it is possible to write

$$\sigma_{ij}\,\delta\epsilon_{ij} = C_{ijkl}\,\epsilon_{ij}\,\delta\epsilon_{ij} = \epsilon_{ij}\,C_{ijkl}\,\delta\epsilon_{ij} \qquad (4.19)$$

and as C_{ijkl} is constant

$$\sigma_{ij}\,\delta\epsilon_{ij} = \epsilon_{ij}\,\delta\sigma_{ij} \qquad (4.20)$$

By introducing this last expression into $\delta\Pi_2$ it reduces to

$$\delta\Pi_2(u_i,\tilde{u}_i,\lambda_i) = \int_\Omega (\sigma_{ij}\,\delta\epsilon_{ij} - b_i\,\delta u_i)\,d\Omega - \int_\Gamma \lambda_i\,\delta u_i\,d\Gamma +$$
$$\int_\Gamma (\tilde{u}_i - u_i)\,\delta\lambda_i\,d\Gamma + \int_{\Gamma_p} (\lambda_i - \bar{p}_i)\,\delta\tilde{u}_i\,d\Gamma \qquad (4.21)$$

The strain-displacement relations (4.1) allows the following relation to be written:

$$\sigma_{ij}\,\delta\epsilon_{ij} = \frac{1}{2}\,\sigma_{ij}\left[\delta\left(\frac{\partial u_i}{\partial x_j}\right) + \delta\left(\frac{\partial u_i}{\partial x_j}\right)\right] \qquad (4.22)$$

From the variational calculus the derivative and the variation symbols can be interchanged, i. e.,

$$\delta\left(\frac{\partial u_i}{\partial x_j}\right) = \frac{\partial}{\partial x_j}(\delta u_i) \tag{4.23}$$

which taken into equation (4.22) produces

$$\sigma_{ij}\,\delta\epsilon_{ij} = \frac{1}{2}\,\sigma_{ij}\left[\delta\left(\frac{\partial u_i}{\partial x_j}\right) + \delta\left(\frac{\partial u_j}{\partial x_i}\right)\right] \tag{4.24}$$

In the last expression, the symmetry of the stress tensor (4.3) leads to

$$\sigma_{ij}\,\delta\epsilon_{ij} = \sigma_{ij}\,\frac{\partial}{\partial x_j}(\delta u_i) \tag{4.25}$$

The introduction of equation (4.25) into expression (4.21) yields:

$$\delta\Pi_2(u_i, \tilde{u}_i, \lambda_i) = \int_\Omega \left[\sigma_{ij}\,\frac{\partial}{\partial x_j}(\delta u_i) - b_i\,\delta u_i\right]d\Omega - \int_\Gamma \lambda_i\,\delta u_i\,d\Gamma +$$
$$\int_\Gamma (\tilde{u}_i - u_i)\,\delta\lambda_i\,d\Gamma + \int_{\Gamma_p} (\lambda_i - \overline{p}_i)\,\delta\tilde{u}_i\,d\Gamma \tag{4.26}$$

The first term on the right-hand side of the above equation, when integrated by parts results in

$$\int_\Omega \sigma_{ij}\,\frac{\partial}{\partial x_j}(\delta u_i)\,d\Omega = \int_\Gamma \sigma_{ij}\,n_j\,\delta u_i\,d\Gamma - \int_\Omega \sigma_{ij,j}\,\delta u_i\,d\Omega \tag{4.27}$$

and by substituting equation (4.4) into the boundary integral in the last expression one obtains

$$\int_\Omega \sigma_{ij}\,\frac{\partial}{\partial x_j}(\delta u_i)\,d\Omega = \int_\Gamma p_i\,\delta u_i\,d\Gamma - \int_\Omega \sigma_{ij,j}\,\delta u_i\,d\Omega \tag{4.28}$$

If equation (4.28) is now taken into consideration in the expression for $\delta\Pi_2$, as given by equation (4.26), and the terms are rearranged, it is possible to write

$$\delta\Pi_2(u_i, \tilde{u}_i, \lambda_i) = -\int_\Omega (\sigma_{ij,j} + b_i)\,\delta u_i\,d\Omega + \int_\Gamma (\tilde{u}_i - u_i)\,\delta\lambda_i\,d\Gamma +$$
$$\int_\Gamma (p_i - \lambda_i)\,\delta u_i\,d\Gamma + \int_{\Gamma_p} (\lambda_i - \overline{p}_i)\,\delta\tilde{u}_i\,d\Gamma \tag{4.29}$$

The Euler equations for the functional are obtained when this first variation is set equal to zero for any values of δu_i, $\delta \tilde{u}_i$ and $\delta \lambda_i$. This means that the stationary conditions for the functional are given by the following equations:

$$\sigma_{ij,j} + b_i \;=\; 0 \;\text{ in } \Omega \tag{4.30}$$

$$\tilde{u}_i - u_i \;=\; 0 \;\text{ on } \Gamma \tag{4.31}$$

$$p_i - \lambda_i \;=\; 0 \;\text{ on } \Gamma \tag{4.32}$$

$$\lambda_i - \overline{p}_i \;=\; 0 \;\text{ on } \Gamma \tag{4.33}$$

From the last two expressions it is possible to conclude that the Lagrange multipliers λ_i represent the tractions on the boundary, which will therefore be denoted by $\tilde{p}_i$.

It is seen that these Euler equations, together with the a priori assumed satisfied equations (4.1), (4.4), (4.5) and (4.16), are equivalent to the basic relations shown in section 4.2, hence they uniquely define the problem.

On the boundary the condition of compatibility of displacements, the traction equilibrium and the natural boundary condition are satisfied in a variational sense as one can see from equations (4.31), (4.32) and (4.33). Consequently the modified variation principle presented completely defines the elasticity problem and can be used to solve it.

Now the Lagrange multipliers will be substituted by the equivalent physical variables — $\tilde{p}_i$ — in the expression for the functional Π_2 in (4.15), which can then be recast into the following form:

$$\Pi_2\left(u_i, \tilde{u}_i, \tilde{p}_i\right) = \int_{\Omega} \left(\frac{1}{2}\,\sigma_{ij}\,\epsilon_{ij} - b_i\,u_i\right) d\Omega + \int_{\Gamma} \tilde{p}_i\,(\tilde{u}_i - u_i)\,d\Gamma - \int_{\Gamma_p} \overline{p}_i\,\tilde{u}_i\,d\Gamma \tag{4.34}$$

Following the same procedures used to demonstrate expression (4.25) it is straightforward to prove that

$$\sigma_{ij}\,\epsilon_{ij} = \sigma_{ij}\,u_{i,j} \tag{4.35}$$

and by substituting this last result into equation (4.34) one has

$$\Pi_2(u_i,\tilde{u}_i,\tilde{p}_i) = \int_\Omega \left(\frac{1}{2}\,\sigma_{ij}\,u_{i,j} - b_i\,u_i\right)d\Omega + \int_\Gamma \tilde{p}_i\,(\tilde{u}_i - u_i)\,d\Gamma - \int_{\Gamma_p} \bar{p}_i\,\tilde{u}_i\,d\Gamma \tag{4.36}$$

The first term on the right-hand-side can be integrated by parts to become

$$\int_\Omega \frac{1}{2}\,\sigma_{ij}\,u_{i,j}\,d\Omega = \int_\Gamma \frac{1}{2}\,\sigma_{ij}\,n_j\,u_i\,d\Gamma - \int_\Omega \frac{1}{2}\,\sigma_{ij,j}\,u_i\,d\Omega \tag{4.37}$$

and finally the substitution of equation (4.4) into that integral allows it to be transformed as follows:

$$\int_\Omega \frac{1}{2}\,\sigma_{ij}\,u_{i,j}\,d\Omega = \int_\Gamma \frac{1}{2}\,p_i\,u_i\,d\Gamma - \int_\Omega \frac{1}{2}\,\sigma_{ij,j}\,u_i\,d\Omega \tag{4.38}$$

which taken into consideration in equation (4.36) produces the final form for the generalized functional Π_{HE}, i.e.,

$$\Pi_{HE}(u_i,\tilde{u}_i,\tilde{p}_i) = \int_\Gamma \frac{1}{2}\,p_i\,u_i\,d\Gamma + \int_\Gamma \tilde{p}_i\,(\tilde{u}_i - u_i)\,d\Gamma - \int_{\Gamma_p} \bar{p}_i\,\tilde{u}_i\,d\Gamma - \int_\Omega b_i\,u_i\,d\Omega - \int_\Omega \frac{1}{2}\,\sigma_{ij,j}\,u_i\,d\Omega \tag{4.39}$$

The ad hoc multi-field variational principle, basis of the proposed formulation, can then be stated as:

Among all the functions u_i, which satisfy equations (4.1), (4.4) and (4.5) and $\tilde{u}_i$, which fulfills the essential boundary condition (4.16), and $\tilde{p}_i$, piecewise continuous function, the actual solution is given by those that make stationary the functional Π_{HE}, defined in equation (4.39).

4.4 Derivation of the Model

This section is concerned with the generation of the final algebraic system of equations which produces an approximate solution for the problem. This is achieved by representing the three independent field variables — displacement in the domain, displacement and traction on the boundary — through approximate functions and then applying the generalized variational principle defined in the previous section.

The approximation for the boundary variables is made by using interpolation functions of the type normally adopted in finite or boundary elements theories. The displacement field in the domain, which is the only independent domain variable, instead, is approximated in a different way: the so-called fundamental solution, extensively used in the boundary element method, is applied here to generate the approximate internal displacement field, as it will be shown in section (4.4.2).

In the following sections, when it is convenient, the matrix notation will be used instead of the indicial notation.

A review of the definitions and terminology used in the classical boundary element formulation [64, 69, 65], which is also used in this work, will be presented in the next section.

4.4.1 Fundamental Solution

The solution of Navier's equation corresponding to Kelvin's problem of a point load in a homogeneous, isotropic, linear, elastic and infinite body is called the Kelvin fundamental solution or simply the fundamental solution of elastostatics. There are other fundamental solutions depending on the type of domain under consideration, such as the half-space fundamental solution (presented by Melan [71]) and the half-plane fundamental solution (due to Midlin [72]) and also the fundamental solution for axisymmetric problems. Nevertheless, the Kelvin solution is the most general one and is the starting point for the derivation of all the others. These solutions are always singular at the point where the load is applied which is characteristic of a concentrated load.

Accordingly, the Kelvin fundamental solution satisfies the Navier equations corresponding to a unit concentrated load, i. e.,

$$G\, u^*_{j,kk} + \frac{G}{1-2\nu}\, u^*_{k,kj} + \Delta(\xi,x)\, e_j = 0 \qquad (4.40)$$

where u^*_j is the j component of the fundamental displacement at point x; $\Delta(\xi,x)$ represents the Dirac delta function (equation 2.48); ξ is the source point and x is the field point whose definitions where given in section 2.5.1 for potential problems and also are applied herein; e_j is the unit vector in the direction of the x_j axis of Cartesian coordinates; the body force $b^* = \Delta(\xi,x)e_j$ is a unit concentrated load applied at point ξ in the j direction; the asterisks refers to fundamental solution.

In the literature, the l components of the fundamental displacement at a point x due to a unit load applied at the source point ξ in k direction are usually denoted by $u^*_{kl}(\xi,x)$ [64, 65]. For Kelvin's problem, they are given by the following expressions:

for the three-dimensional case:

$$u^*_{kl}(\xi,x) = \frac{1}{16\pi G(1-\nu)r}[(3-4\nu)\delta_{kl} + r_{,k}r_{,l}] \qquad (4.41)$$

and the corresponding traction components are

$$p_{kl}^*(\xi, x) = \frac{-1}{8\pi(1-\nu)r^2} \left[\frac{\partial r}{\partial n}[(1-2\nu)\delta_{kl} + 3\,r_{,k}r_{,l}] + (1-2\nu)(n_k r_{,l} - n_l r_{,k}) \right]$$

$$(4.42)$$

in the two-dimensional case of plane strain problems the displacement components are given by

$$u_{kl}^*(\xi, x) = \frac{1}{8\pi G(1-\nu)r} \left[(3-4\nu)\ln(\frac{1}{r})\delta_{kl} + r_{,k}r_{,l} \right] \qquad (4.43)$$

and the related traction component by

$$p_{kl}^*(\xi, x) = \frac{-1}{4\pi(1-\nu)r} \left[\frac{\partial r}{\partial n}[(1-2\nu)\delta_{kl} + 2r_{,k}r_{,l}] + (1-2\nu)(n_k r_{,l} - n_l r_{,k}) \right]$$

$$(4.44)$$

Note that in these expressions k and l represent the orthogonal coordinate directions. Hence k and l have a range of 3 in three-dimensional and 2 in two-dimensional problems.

4.4.2 Approximation for the Domain Variable

The l component of the displacement vector at a point x inside the domain Ω — $u_l(x)$ — will be approximated as a series of products of fundamental solutions and unknown parameters as follows:

$$u_l(x) = u_{1l}^1(x)\,\gamma_1 + u_{2l}^1(x)\,\gamma_2 + u_{1l}^2(x)\,\gamma_3 + u_{2l}^2(x)\,\gamma_4 + \ldots + u_{1l}^N(x)\,\gamma_{2N-1} +$$

$$u_{2l}^N(x)\,\gamma_{2N} \qquad (4.45)$$

or more concisely as:

$$u_l(x) = \sum_{i=1}^{i=N} u_{kl}^{*i}(x)\,\gamma_s \qquad (4.46)$$

where s is given by $s = 2i - 2 + k$; the unit load is applied at node i in the k direction; k and l have a range of 2 or 3, respectively in two or three dimensions. The total number of nodes is denoted by N.

Notice that a slightly different notation has been introduced here for the fundamental solution (see section 4.4.1): the difference corresponds to the use of a superscript to indicate the position of the source point. This has been done for the sake of clarity in the development of the formulation. From now on this notation will be used unless otherwise stated.

Bearing in mind a physical explanation for this type of approximation, the parameters γ_s can be understood as fictitious concentrated loads in an infinite domain at points which correspond to boundary points in the actual body. Both domains, the actual and the infinite domain, should have the same elastic properties. The displacement field in the actual domain Ω is then approximated as the displacement field in the infinite domain Ω_∞ subject to the action of all these fictitious loads.

Equation (4.46) can be represented in matrix notation as

$$\mathbf{u} = \mathbf{U}^{*T}\,\boldsymbol{\gamma} \qquad (4.47)$$

where the displacement vector at a point in the domain, $\mathbf{u}$ has the form

$$\mathbf{u} = \left\{ \begin{array}{c} u_1 \\ u_2 \\ u_3 \end{array} \right\} \qquad (4.48)$$

the vector of fictitious loads $\boldsymbol{\gamma}$ is defined as

$$\boldsymbol{\gamma} = \left\{ \begin{array}{c} \gamma_1 \\ \gamma_2 \\ \vdots \\ \gamma_{3N} \end{array} \right\} \qquad \text{(three-dimensions)} \qquad (4.49)$$

and the matrix of fundamental solutions $\mathbf{U}^*$ is given by

$$\mathbf{U}^* = \begin{bmatrix} u_{11}^{*1} & u_{12}^{*1} & u_{13}^{*1} \\[2pt] u_{21}^{*1} & u_{22}^{*1} & u_{23}^{*1} \\[2pt] u_{31}^{*1} & u_{32}^{*1} & u_{33}^{*1} \\[2pt] u_{11}^{*2} & u_{12}^{*2} & u_{13}^{*2} \\[2pt] u_{21}^{*2} & u_{22}^{*2} & u_{23}^{*2} \\[2pt] u_{31}^{*2} & u_{32}^{*2} & u_{33}^{*2} \\[2pt] \vdots & \vdots & \vdots \\[2pt] u_{11}^{*N} & u_{12}^{*N} & u_{13}^{*N} \\[2pt] u_{21}^{*N} & u_{22}^{*N} & u_{23}^{*N} \\[2pt] u_{31}^{*N} & u_{32}^{*N} & u_{33}^{*N} \end{bmatrix} \qquad \text{(three-dimensions)} \qquad (4.50)$$

The matrix of tractions $\mathbf{P}^*$ associated with the matrix of the fundamental solutions $\mathbf{U}^*$ is represented as shown below:

$$\mathbf{P}^* = \begin{bmatrix} p_{11}^{*1} & p_{12}^{*1} & p_{13}^{*1} \\[2pt] p_{21}^{*1} & p_{22}^{*1} & p_{23}^{*1} \\[2pt] p_{31}^{*1} & p_{32}^{*1} & p_{33}^{*1} \\[2pt] p_{11}^{*2} & p_{12}^{*2} & p_{13}^{*2} \\[2pt] p_{21}^{*2} & p_{22}^{*2} & p_{23}^{*2} \\[2pt] p_{31}^{*2} & p_{32}^{*2} & p_{33}^{*2} \\[2pt] \vdots & \vdots & \vdots \\[2pt] p_{11}^{*N} & p_{12}^{*N} & p_{13}^{*N} \\[2pt] p_{21}^{*N} & p_{22}^{*N} & p_{23}^{*N} \\[2pt] p_{31}^{*N} & p_{32}^{*N} & p_{33}^{*N} \end{bmatrix} \qquad \text{(three-dimensions)} \qquad (4.51)$$

where the indices have the same meaning they have in the components of matrix $\mathbf{U}^*$.

In two-dimensional problems $\mathbf{U}^*$ and $\mathbf{P}^*$ are $2N \times 2$ rectangular matrices, instead of $3N \times 3$, and Γ has $2N$ rows.

4.4.3   Approximation for Boundary Variables

The boundary displacements and tractions vectors denoted by $\tilde{u}$ and $\tilde{p}$ are represented in terms of the values of those variables at a series of boundary nodes. They are written therefore as the product of known interpolation functions by unknown parameters, i.e.,

$$\tilde{\mathbf{u}}(x) = \mathbf{\Phi}^T \mathbf{u} \qquad x \, \varepsilon \, \Gamma \qquad (4.52)$$

$$\tilde{\mathbf{p}}(x) = \mathbf{\Psi}^T \mathbf{p} \qquad x \, \varepsilon \, \Gamma \qquad (4.53)$$

where the coefficients of the matrices $\mathbf{\Phi}$ and $\mathbf{\Psi}$ are interpolation functions of the type normally adopted by the finite or the boundary element methods and T denotes transpose; $\mathbf{u}$ and $\mathbf{p}$ are vectors whose components are nodal values for boundary displacements and boundary tractions, respectively, and are written as

$$\mathbf{u} = \left\{ \begin{array}{c} u_1 \\ u_2 \\ \vdots \\ u_N \end{array} \right\} \qquad (4.54)$$

$$\mathbf{p} = \left\{ \begin{array}{c} p_1 \\ p_2 \\ \vdots \\ p_N \end{array} \right\} \qquad (4.55)$$

Observe that the vectors $\mathbf{u}$ and $\mathbf{p}$ represent nodal values of the boundary variables $\tilde{u}$ and $\tilde{p}$ although the tilde has been dropped from the notation for simplicity.

4.4.4 Final System of Equations

The final system of algebraic equations is obtained as the stationary conditions of the functional Π_{HE} (4.39), after introducing into its expression the approximate field variables u, $\tilde{u}$ and $\tilde{p}$, as defined in equations (4.47), (4.52) and (4.53), respectively.

As the functions u^*_{kl} and p^*_{kl} are singular at the boundary nodes a new domain Ω' is defined from Ω by excluding the singular points, as shown in figure 4.2. Parts of small spheres (circles in two-dimensions) of radius ϵ, centred at the boundary nodes, are then removed from Ω to generate the domain Ω'. Consequently in the limit, when ϵ tends to zero, the domain Ω' also tends to Ω including all its boundaries, i. e.,

when $\varepsilon \longrightarrow 0$ then:

$$\Omega' \longrightarrow \Omega \,, \qquad \Gamma' \longrightarrow \Gamma \,, \qquad \Gamma'_u \longrightarrow \Gamma_u \,, \qquad \Gamma'_p \longrightarrow \Gamma_p$$

where Γ', Γ'_u and Γ'_p are the boundaries in the new domain Ω' corresponding to the original boundaries, Γ, Γ_u and Γ_p defined in Ω.

By applying these limit considerations it is possible to write the functional (4.39) as

$$\Pi_{HE}(u_i,\tilde{u}_i,\tilde{p}_i) \;=\; \lim_{\varepsilon \to 0} \left\{ \int_{\Gamma'} \frac{1}{2}\, p_i\, u_i\, d\Gamma + \int_{\Gamma'} \tilde{p}_i\, (\tilde{u}_i - u_i)\, d\Gamma - \right.$$
$$\left. \int_{\Gamma'_p} \bar{p}_i\, \tilde{u}_i\, d\Gamma - \int_{\Omega'} b_i\, u_i\, d\Omega - \int_{\Omega'} \frac{1}{2}\, \sigma_{ij,j}\, u_i\, d\Omega \right\} \tag{4.56}$$

Now the approximate variables u, $\tilde{u}$ and $\tilde{p}$, from equations (4.47), (4.52) and (4.53), are introduced into the last expression. This yields

$$\Pi_{HE} \;=\; \lim_{\varepsilon \to 0} \left\{ \int_{\Gamma'} \frac{1}{2}\, \gamma^T\, \mathbf{U}^*\, \mathbf{P}^{*T}\, \gamma\, d\Gamma + \int_{\Gamma'} \mathbf{p}^T\, \boldsymbol{\Psi}\, (\boldsymbol{\Phi}^T\, \mathbf{u} - \mathbf{U}^{*T}\, \gamma)\, d\Gamma - \right.$$
$$\left. \int_{\Gamma'_p} \mathbf{u}^T\, \boldsymbol{\Phi}\, \bar{\mathbf{p}}\, d\Gamma - \int_{\Omega'} \gamma^T\, \mathbf{U}^*\, \mathbf{b}\, d\Omega \right\} \tag{4.57}$$

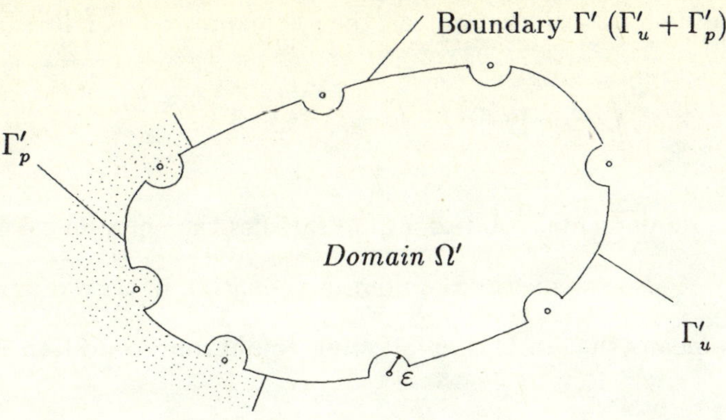

Figure 4.2: Domain Ω' and boundary Γ'

where $\overline{\mathbf{p}}$ is the vector of prescribed tractions at a point on the boundary and $\mathbf{b}$ is the vector of body forces at a point in the domain.

The last integral in equation (4.56) vanishes due to the type of approximation used for the displacement in the domain as it will be explained in what follows.

As a consequence of the approximation used for the internal displacement (4.47), the stress tensor at an internal point x — $\sigma_{ij}(x)$ — can be written in terms of the stress tensor at x in the infinite domain when subject to the action of the fictitious sources γ_k $(k = 1, 2, ..., 2N)$. If the stress tensor at x due to one component γ_k is denoted by $\sigma_{ij}^{*k}(x)$ then one has

$$\sigma_{ij}(x) = \sigma_{ij}^{*k}(x)\,\gamma_k \tag{4.58}$$

in which $k = 1, 2, .., 2N$ and the summation convention is implied in relation to the k index.

The integral under consideration can then be expressed as follows:

$$\int_{\Omega'} \sigma_{ij,j}\, u_i\, d\Omega = \int_{\Omega'} \sigma^{*k}_{ij,j}\, \gamma_k\, u_i\, d\Omega \qquad (4.59)$$

Because the fundamental solution $u^*_{kl}(x)$ satisfies the equation (4.40) in the domain Ω_∞ it also satisfies the equilibrium equation in term of stresses (4.2), and this means that in Ω' the following equilibrium condition is fulfilled:

$$\sigma^{*k}_{ij,j} = 0 \qquad \text{in } \Omega' \qquad (4.60)$$

because the load γ_k acts outside the domain Ω'.

Finally by taking this result into equation (4.59) it becomes

$$\int_{\Omega'} \sigma_{ij,j}\, u_i\, d\Omega = 0 \qquad (4.61)$$

and the value of the integral has been proved to be zero.

If equation (4.57) is rearranged and, in addition to that, if the constant terms are taken out of the integral sign one obtains

$$\Pi_{HE} = \lim_{\varepsilon \to 0} \left\{ \frac{1}{2}\, \gamma^T \int_{\Gamma'} \mathbf{U}^* \mathbf{P}^{*T}\, d\Gamma\, \gamma - \mathbf{p}^T \int_{\Gamma'} \mathbf{\Psi}\, \mathbf{U}^{*T}\, d\Gamma\, \mathbf{u} + \right.$$
$$\left. \mathbf{p}^T \int_{\Gamma'} \mathbf{\Psi}\, \mathbf{\Phi}^T\, d\Gamma\, \mathbf{u} - \mathbf{u}^T \int_{\Gamma'_p} \mathbf{\Phi}\, \overline{\mathbf{p}}\, d\Gamma - \gamma^T \int_{\Omega'} \mathbf{U}^*\, \mathbf{b}\, d\Gamma \right\} (4.62)$$

The following definitions are now introduced:

$$\mathbf{F} = \lim_{\varepsilon \to 0} \int_{\Gamma'} \mathbf{U}^* \mathbf{P}^{*T}\, d\Gamma \qquad (4.63)$$

$$\mathbf{G} = \lim_{\varepsilon \to 0} \int_{\Gamma'} \mathbf{U}^* \mathbf{\Psi}^T\, d\Gamma \qquad (4.64)$$

$$\mathbf{L} = \lim_{\varepsilon \to 0} \int_{\Gamma'} \mathbf{\Psi}\, \mathbf{\Phi}^T\, d\Gamma = \int_{\Gamma} \mathbf{\Psi}\, \mathbf{\Phi}^T\, d\Gamma \qquad (4.65)$$

$$\overline{\mathbf{P}} = \lim_{\varepsilon \to 0} \int_{\Gamma'_2} \mathbf{\Phi}\, \overline{\mathbf{p}}\, d\Gamma = \int_{\Gamma_p} \mathbf{\Phi}\, \overline{\mathbf{p}}\, d\Gamma \qquad (4.66)$$

$$\mathbf{B} = \lim_{\varepsilon \to 0} \int_{\Omega'} \mathbf{U}^*\, \mathbf{b}\, d\Omega \qquad (4.67)$$

In the cases of matrices $\mathbf{L}$ and $\overline{\mathbf{P}}$ there are no singularities and hence it is possible to write, instead of the limit when ε tends to zero of the integral over Γ', simply the integral over Γ, as it has been indicated.

By taking these definitions into the expression for the functional its matrix form results in

$$\Pi_{HE} = \frac{1}{2}\,\gamma^T\,\mathbf{F}\,\gamma - \mathbf{p}^T\,\mathbf{G}^T\,\mathbf{u} + \mathbf{p}^T\,\mathbf{L}\,\mathbf{u} - \mathbf{u}^T\,\overline{\mathbf{P}} - \gamma^T\,\mathbf{B} \qquad (4.68)$$

The first variation for Π_{HE} will now be obtained by taking variations in equation (4.68) with respect to the unknown parameters which are the components of the vectors γ, $\mathbf{u}$ and $\mathbf{p}$, i. e.,

$$\delta\Pi_{HE} = \frac{1}{2}(\delta\gamma)^T\mathbf{F}\,\gamma + \frac{1}{2}\gamma^T\mathbf{F}\,\delta\gamma - (\delta\mathbf{p})^T\mathbf{G}^T\gamma - \mathbf{p}^T\mathbf{G}^T\delta\gamma +$$
$$(\delta\mathbf{p})^T\mathbf{L}\,\mathbf{u} + \mathbf{p}^T\mathbf{L}\,\delta\mathbf{u} - (\delta\mathbf{u})^T\overline{\mathbf{P}} - (\delta\gamma)^T\mathbf{B} \qquad (4.69)$$

The functional Π_{HE} is a scalar and so are all the terms in the last expression. This property can be used together with the symmetry of the matrix $\mathbf{F}$ (see proof in section 4.4.7) to recast $\delta\Pi_{HE}$ as

$$\delta\Pi_{HE} = (\delta\gamma)^T\,(\mathbf{F}\,\gamma - \mathbf{G}\,\mathbf{p} - \mathbf{B}) + \delta\mathbf{p}^T\,(-\mathbf{G}^T\gamma + \mathbf{L}\,\mathbf{u}) +$$
$$(\delta\mathbf{u})^T\,(\mathbf{L}^T\mathbf{p} - \overline{\mathbf{P}}) \qquad (4.70)$$

The stationary conditions for Π_{HE} can now be found by setting to zero its first variation. As this must be true for any arbitrary values of $\delta\gamma$, $\delta\mathbf{u}$ and $\delta\mathbf{p}$ one obtains the three following matrix equations:

$$\mathbf{F}\,\gamma - \mathbf{G}\,\mathbf{p} - \mathbf{B} = 0 \qquad (4.71)$$

$$-\mathbf{G}^T\gamma + \mathbf{L}\,\mathbf{u} = 0 \qquad (4.72)$$

$$\mathbf{L}^T\mathbf{p} - \overline{\mathbf{P}} = 0 \qquad (4.73)$$

The approximate solution of the problem is the solution of the above system of equations because it makes the functional Π_{HE} stationary.

Now the unknowns γ and $\mathbf{p}$ will be expressed in terms of $\mathbf{u}$ to obtain a final equation involving only the boundary unknown $\mathbf{u}$, as follows:

The matrix $\mathbf{G}^T$ is a non-singular square matrix, hence it is invertible and γ can be obtained from equation (4.72) as

$$\gamma = \left(\mathbf{G}^T\right)^{-1} \mathbf{L}\,\mathbf{u} \qquad (4.74)$$

which taken to equation (4.71) produces a expression for $\mathbf{p}$, i. e.,

$$\mathbf{p} = \mathbf{G}^{-1}\,\mathbf{F}\,\left(\mathbf{G}^T\right)^{-1}\,\mathbf{L}\,\mathbf{u} - \mathbf{G}^{-1}\,\mathbf{B} \qquad (4.75)$$

The final matrix equation for the solution of the problem is found by replacing γ and $\mathbf{p}$ from equations (4.74) and (4.75) into (4.73) as shown:

$$\mathbf{K}\,\mathbf{u} - \mathbf{Q} = 0 \qquad (4.76)$$

where

$$\mathbf{K} \;=\; \mathbf{R}^T\mathbf{F}\,\mathbf{R} \qquad (4.77)$$

$$\mathbf{R} \;=\; \left(\mathbf{G}^T\right)^{-1}\mathbf{L} \qquad (4.78)$$

$$\mathbf{Q} \;=\; \overline{\mathbf{P}} + \mathbf{R}^T\mathbf{B} \qquad (4.79)$$

It is possible to conclude from equation (4.76) that this hybrid displacement boundary formulation leads to an equivalent stiffness approach: the matrix $\mathbf{K}$ may be viewed as a symmetric boundary stiffness matrix and the vector $\mathbf{Q}$ as a consistent load vector.

The symmetry of the matrix $\mathbf{K}$ will be proved in section 4.4.7.

Notice that the unknown **u** is defined on the boundary only and the stiffness matrix **K** is computed by performing exclusively boundary integrations.

The other primary unknowns — γ and **p** — can be computed from the boundary displacement **u**.

4.4.5 Solution on the Boundary

Boundary Displacements:

The solution of the system of algebraic equations (4.76) produces the vector of nodal displacements, **u**.

Boundary Tractions:

The tractions at the boundary nodes can be evaluated by taking the solution for boundary displacements into equation (4.75), i. e.,

$$\mathbf{p} = \mathbf{G}^{-1}\,\mathbf{F}\,(\mathbf{G}^{-1})^{T}\,\mathbf{L}\,\mathbf{u} - \mathbf{G}^{-1}\,\mathbf{B} \tag{4.80}$$

4.4.6 Solution at Internal Points

Displacements at Internal Points:

After having solved equation (4.76) the vector γ can be obtained from (4.74) and the l component of the displacement vector at a point x inside the domain Ω — $u_l^{int}(x)$ — can be computed by using expressions (4.46) as follows:

$$u_l^{int}(x) = u_{kl}^{*}(x)\,\gamma_k \qquad x\,\varepsilon\,\Omega \tag{4.81}$$

Stresses at Internal Points:

A component of the stress tension at an internal point x — $\sigma_{ij}^{int}(x)$ — can be found through equation (4.58) which is repeated here for completeness:

$$\sigma_{ij}(x) = \sigma_{ij}^{*k}(x)\, \gamma_k \qquad x\, \varepsilon\, \Omega \qquad (4.82)$$

4.4.7 Symmetry of the Stiffness Matrix

One of the convenient features of this hybrid boundary formulation is the symmetry of the resulting stiffness matrix while the stiffness matrices obtained from the classical boundary elements are not symmetric, unless an artificial symmetrization is enforced [64, 73, 80]

The stiffness matrix $\mathbf{K}$ is defined by equation (4.77) which is repeated herein:

$$\mathbf{K} = \mathbf{R}^T\, \mathbf{F}\, \mathbf{R} \qquad (4.83)$$

From this relationship it is possible to state that the symmetry of matrix $\mathbf{F}$ implies the symmetry of $\mathbf{K}$. In consequence of that the efforts now will be concentrated in proving the symmetry of matrix $\mathbf{F}$ which is defined by equation (4.63). Hence an entry of $\mathbf{F}$ — f_{rs} — is given by the following relationship:

$$f_{rs} = \lim_{\varepsilon \to 0} \int_{\Gamma'} u_{km}^{*i}\, p_{lm}^{*j}\, d\Gamma \qquad (4.84)$$

where

$$r = 2i - 2 + k \qquad (4.85)$$

$$s = 2j - 2 + l \qquad (4.86)$$

and u_{km}^{*i} is the fundamental displacement in m direction due to a unit load in the k direction applied at the node i. Similar notation is used for p_{lm}^{*j}; $k, l, m = 1, 2$ for two-dimensional and $k, l, m = 1, 2, 3$ for three-dimensional bodies. Since i and j vary from 1 to N, r and s have a range of $2N$ or $3N$ in 2-D or 3-D problems, respectively.

In order to prove the symmetry of the matrix $\mathbf{F}$ two different loading conditions should be considered for the same infinite body. The first one, here called loading i, corresponds to a concentrated unit load acting at the boundary Γ at the point i in the k coordinate direction. In the second loading, designed as loading j, the unit load acts in the l direction at the boundary node j.

The basic equations in section 4.2 are valid for these situations. Thus, as elastic properties are the same for both cases, the symmetry of the elastic constants tensor implies :

$$\sigma_{mq}^{*r} \, \epsilon_{mq}^{*s} = \sigma_{mq}^{*s} \, \epsilon_{mq}^{*r} \tag{4.87}$$

where the superscripts r and s are defined in equations (4.85) and (4.86). For loading i the strain-displacement relationship (4.1), according to the current notation, is written as

$$\epsilon_{mq}^{*r} = \frac{1}{2} \left(\frac{\partial u_{km}^{*i}}{\partial x_q} + \frac{\partial u_{kq}^{*i}}{\partial x_m} \right) \tag{4.88}$$

and similarly for the loading j

$$\epsilon_{mq}^{*s} = \frac{1}{2} \left(\frac{\partial u_{lm}^{*j}}{\partial x_q} + \frac{\partial u_{lq}^{*j}}{\partial x_m} \right) \tag{4.89}$$

Since the stress tensor is symmetric, if the last two expressions are brought into equation (4.87) this can be written as

$$\sigma_{mq}^{*r} \frac{\partial u_{lm}^{*j}}{\partial x_q} = \sigma_{mq}^{*s} \frac{\partial u_{km}^{*i}}{\partial x_q} \tag{4.90}$$

This equation is then integrated over the domain Ω' and both terms integrated by parts to produce:

$$\int_{\Gamma'} \sigma_{mq}^{*r} \, u_{lm}^{*j} \, n_q \, d\Gamma - \int_{\Omega'} \sigma_{mq,q}^{*r} \, u_{lm}^{*j} \, d\Omega = \int_{\Gamma'} \sigma_{mq}^{*s} \, u_{km}^{i} \, n_q \, d\Gamma - \int_{\Omega'} \sigma_{mq,q}^{*s} \, u_{km}^{*i} \, d\Omega$$

$$(4.91)$$

According to the definition of Kelvin's fundamental solution and because the domain Ω' does not include the source point, for both loadings, it is possible to state that inside the domain Ω' the following relations hold:

$$\sigma_{mq,q}^{*r} = 0 \quad \text{and} \quad \sigma_{mq,q}^{*s} = 0 \tag{4.92}$$

By substituting these results into expression (4.91) and using equation (4.4) one obtains

$$\int_{\Gamma'} p_{km}^{*i} \, u_{lm}^{*j} \, d\Gamma = \int_{\Gamma'} p_{lm}^{*j} \, u_{km}^{*i} \, d\Gamma \tag{4.93}$$

Which taken to the limit when ε tends to zero results in

$$\lim_{\varepsilon \to 0} \int_{\Gamma'} p_{km}^{*i} \, u_{lm}^{*j} \, d\Gamma = \lim_{\varepsilon \to 0} \int_{\Gamma'} p_{lm}^{*j} \, u_{km}^{*i} \, d\Gamma \tag{4.94}$$

This last expression, according to equation (4.84), means that matrix $\mathbf{F}$ is symmetric, i.e.,

$$f_{sr} = f_{rs} \tag{4.95}$$

Finally it is possible to conclude that the stiffness matrix $\mathbf{K}$ is symmetric once the symmetry of $\mathbf{F}$ has been proved.

Chapter 5

Numerical Aspects in Elastostatics Problems

5.1 Introduction

This chapter is concerned with the computer implementation of the hybrid displacement boundary element formulation presented in chapter 4. The case of two-dimensional problems will be considered, for simplicity. Similar considerations could be used in the implementation of the formulation for the three-dimensional case, although some additional work would be required. Special attention is dedicated to the treatment of some singular integrals which appear in the generation of the matrices $\mathbf{F}$ and $\mathbf{G}$.

In order to obtain the final algebraic system of equations, which will produce the solution of the problem, the matrices $\mathbf{F}$, $\mathbf{G}$, $\mathbf{L}$ and the vectors $\overline{\mathbf{P}}$ and $\mathbf{B}$, defined in chapter 4, have to be evaluated. The definitions of those matrices and vectors (except vector $\mathbf{B}$) involve boundary integrals.

For their numerical evaluation, the boundary needs then to be subdivided into elements.

Over each of these elements, the independent boundary variables — boundary displacements and tractions — are approximated in terms of interpolation functions and nodal values of these variables. This results in the kind of approximation shown in equations (4.52) and (4.53).

The nodes are points over the element, which are usually chosen to be located at particular positions, such as the midpoint or the end points of the element. The number of nodes per element depends on the degree of the interpolation functions adopted. For example; one, two or three nodes have to be defined if the interpolation functions are constant, linear or quadratic, respectively.

In this work, two types of approximation are used for the boundary variables. The first corresponds to the situation in which the displacements and the tractions are assumed to be constant along the element. In this case, the geometry of the element is approximated by a straight line. For the second type of approximation, the geometry and the two boundary variables are supposed to vary quadratically, resulting in the need for the definition of three nodes per element. These are chosen to be at the end points and at the midpoint of the element. The resulting element is accordingly called isoparametric quadratic element, following the well established terminology adopted by other numerical techniques (finite or conventional boundary elements).

The singular integrals in the terms of matrices $\mathbf{F}$ and $\mathbf{G}$ are treated either by analytical procedures or by special numerical integration schemes. In elastostatics problems, as in the potential case, the matrix $\mathbf{F}$ does not vary with the type of approximation for the boundary variables. It depends only on the fundamental solution adopted and on the geometry of the element. The matrix $\mathbf{G}$ in this approach is similar to the one that appears in the conventional direct boundary element method formulation. Therefore its evaluation, including the treatment of the singular integrals, follows the same lines as in the conventional boundary elements.

The elements of matrix $\mathbf{L}$ and the vector $\overline{\mathbf{P}}$ do not include any kind of singularity. For the constant element, they are simple to obtain directly, without the need for any type of integration. In the case of the quadratic element, some simple boundary integrations have to be carried out. A Gauss-Legendre quadrature formula may be applied for their computation.

In this formulation, as well as in the classical boundary element method, the computation of domain integrals involves a major drawback. In principle, the computation of such integrals requires a domain discretization, which demands a more elaborated input data. Some boundary element researchers [77, 78, 84, 79] have studied the possibility of taking these domain integrals to the boundary, in such a way that the evaluation of integrals over the domain can be avoided.

When body forces are present in the problem, the domain integrals corresponding to the coefficients of the vector $\mathbf{B}$ are given by the same domain integrals that have to be computed in the classical direct boundary element method. One can then take advantage of this fact and proceed as in the conventional boundary elements, when considering the body force integrals. The computation of matrix $\mathbf{B}$ is explained is section 5.7.

In this hybrid-displacement approach, the evaluation of the internal displacements and stresses, when body forces are present, does not require the computation of any volume integral. In this case, internal displacements and stresses are then computed by applying the same expressions used in the case of zero body forces (chapter 4, equations (4.46) and (4.58)).

Throughout this chapter the plane strain case is considered but plane stress problems can also be solved using the formulae presented. In order to do so it is only necessary to replace the elastic constants, Young's modulus E and Poisson's ratio ν, by two equivalent values E' and ν', which are given by

$$E' = (1 - \nu^2)\, E \qquad\qquad \nu' = \frac{\nu}{1 + \nu} \tag{5.1}$$

5.2   The Constant Element

This type of element can be used in problems whose solution along the boundary has a smooth variation. The problems should also have boundaries that can be reasonably approximated by straight lines.

The boundary Γ is discretized into a series of straight elements where the displacements and tractions are assumed to be constant , i.e.

$$\mathbf{u} \;=\; \mathbf{u}^k \qquad \text{on the element } \Gamma_k \tag{5.2}$$

$$\mathbf{p} \;=\; \mathbf{p}^k \qquad \text{on the element } \Gamma_k \tag{5.3}$$

which are an specialization of equations (4.52) and (4.53) for the element Γ_k; k varies from 1 to N; N is the number of boundary nodes.

The element's geometry is straight and defined by its end points herein called point(1) and point(2). See figure 5.1.

The homogeneous coordinate η, defined in section 3.2 for potential problems is shown in figure 5.1. It will be used in the evaluation of the integrals which gives the elements of matrices $\mathbf{F}$ and $\mathbf{G}$.

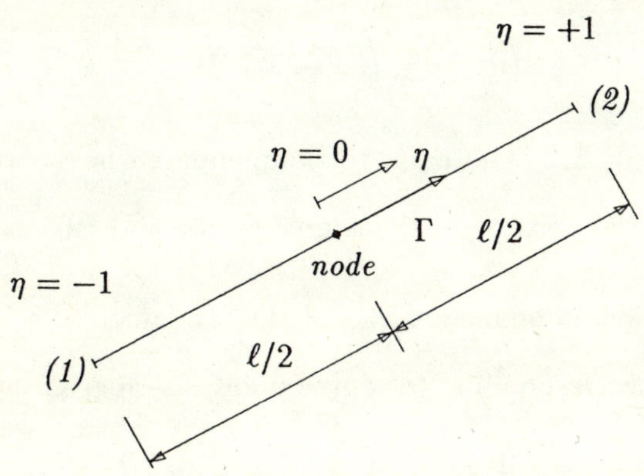

Figure 5.1: Constant element

5.2.1   Matrix F for Constant Elements

The matrix $\mathbf{F}$ is a $2N \times 2N$ square matrix and is defined in equation (4.63). It can also be written as a partitioned matrix, as shown below:

$$\mathbf{F} = \begin{bmatrix} \mathcal{F}^{11} & \mathcal{F}^{12} & \cdots & \mathcal{F}^{1N} \\ \mathcal{F}^{21} & \mathcal{F}^{22} & \cdots & \mathcal{F}^{2N} \\ \vdots & \vdots & \vdots & \vdots \\ \mathcal{F}^{N1} & \mathcal{F}^{N2} & \cdots & \mathcal{F}^{NN} \end{bmatrix} \tag{5.4}$$

where $\mathcal{F}^{ij}$ is a 2×2 submatrix that corresponds to a particular position of the source points i and j. The range for i and j is equal the total number of boundary nodes, N.

The submatrices $\mathcal{F}^{ij}$ are defined as follows:

$$\mathcal{F}^{ij} = \begin{bmatrix} f_{11}^{ij} & f_{12}^{ij} \\ f_{21}^{ij} & f_{22}^{ij} \end{bmatrix} \tag{5.5}$$

and a generic entry of $\mathcal{F}^{ij}$ — f_{kl}^{ij} — is expressed as

$$f_{kl}^{ij} = \lim_{\varepsilon \to 0} \int_{\Gamma'} u_{km}^{*i} \, p_{lm}^{*j} \, d\Gamma \tag{5.6}$$

The indices $k, l, m = 1, 2$ denote matrix components whereas the subscripts i and j represent the position of the source points and have a range of N.

The displacement components u_{km}^{*i} of the fundamental solution for the plain strain case, corresponding to source point i, are given by

$$u_{km}^{*i} = \frac{1}{8 \, \pi \, E \, (1 - \nu) \, r^i} \left[(3 - 4\nu) \, \delta_{km} + \frac{\partial r^i}{\partial x_k} \frac{\partial r^i}{\partial x_m} \right] \tag{5.7}$$

and the related traction components, when the source point is at j, are

$$p_{km}^{*j} = \frac{-1}{4 \, \pi (1 - \nu) \, r^i} \left[\frac{\partial r^j}{\partial n} (1 - 2\,\nu) \, \delta_{km} + 2 \frac{\partial r^j}{\partial x_k} \frac{\partial r^j}{\partial x_m} + \right.$$

$$\left. (1 - 2\nu) \left(\frac{\partial r^j}{\partial x_m} n_k - \frac{\partial r^j}{\partial x_k} n_m \right) \right] \tag{5.8}$$

Note that r^i and r^j represent the distances between the field point and the source points which are located at nodes i and j in the expressions for u^{*i} and p^{*j}, respectively.

Expression (5.6) can therefore be written as

$$f_{kl}^{ij} = \lim_{\varepsilon \to 0} \int_{\Gamma'} \frac{C_o}{r^j} \left[(3 - 4v) \, \ln(r^i) \, \delta_k m - \frac{\partial r^i}{\partial x_k} \frac{\partial r^i}{\partial x_m} \right]$$

$$\left\{ \left[(1 - 2v) \, \delta_{lm} + 2 \frac{\partial r^j}{\partial x_l} \frac{\partial r^j}{\partial x_m} \right] \frac{\partial r^j}{\partial n} - (1 - 2v) \left(\frac{\partial r^j}{\partial x_l} n_m - \frac{\partial r^j}{\partial x_m} n_l \right) \right\} \tag{5.9}$$

where C_o is a constant which depends on the elastic constants of the material: Young's modulus E and Poisson's ratio ν, as shown:

$$C_o = \frac{1}{32\pi^2(1-\nu)^2\,E} \tag{5.10}$$

For the discretized boundary equation (5.9) becomes

$$f_{kl}^{ij} = \sum_{e=1}^{N} \lim_{\varepsilon \to 0} \int_{\Gamma'_e} \frac{C_o}{r^j} \left[(3-4v)\,\ln(r^i)\,\delta_k m - \frac{\partial r^i}{\partial x_k}\,\frac{\partial r^i}{\partial x_m} \right]$$

$$\left\{ \left[(1-2v)\,\delta_{lm} + 2\,\frac{\partial r^j}{\partial x_l}\,\frac{\partial r^j}{\partial x_m} \right] \frac{\partial r^j}{\partial n} - (1-2v) \left(\frac{\partial r^j}{\partial x_l}\,n_m - \frac{\partial r^j}{\partial x_m}\,n_l \right) \right\} \tag{5.11}$$

where Γ'_e is the element Γ_e after excluding any singularities that may be present on the element, i.e. Γ'_e is related to the domain Ω'.

Depending on whether or not the sources i and j belong to the element Γ_e, the integrals in (5.11) can be singular. When they are located on the element, three different kinds of singularities occur. Consequently four different situations should be considered in the evaluation of the integrals, these are:

- There are no source points on the element.

- The source point i is located on the element.

- The source point j is located on the element.

- Both source points i and j coincide and are located on the element.

The four situations will now be discussed separately.

Case 1: *Case for which there are no source points on the element under consideration.*

The integral to be computed is regular and therefore can be expressed as

$$\int_{\Gamma_e} \frac{C_o}{r^j} \left[(3 - 4v) \, \ln(r^i) \, \delta_{km} - \frac{\partial r^i}{\partial x_k} \frac{\partial r^i}{\partial x_m} \right]$$

$$\left\{ \left[(1 - 2v) \, \delta_{lm} + 2 \frac{\partial r^j}{\partial x_l} \frac{\partial r^j}{\partial x_m} \right] \frac{\partial r^j}{\partial n} - (1 - 2v) \left(\frac{\partial r^j}{\partial x_l} n_m - \frac{\partial r^j}{\partial x_m} n_l \right) \right\}$$

$$(5.12)$$

Since the element is straight the derivative of r^j with respect to the normal $\mathbf{n}$, given by

$$\frac{\partial r^j}{\partial n} = \frac{\mathbf{r}^j \cdot \mathbf{n}}{r^j} \tag{5.13}$$

is equal to zero, due to the orthogonality between the vectors $\mathbf{r}^j$ and $\mathbf{n}$. The expression (5.12) then becomes

$$\int_{\Gamma_e} \frac{C_o}{r^j} \left[(3 - 4\nu) \ln(r^i) \, \delta_{km} - \frac{\partial r^i}{\partial x_k} \frac{\partial r^i}{\partial x_m} \right]$$

$$\left[(1 - 2\nu) \left(\frac{\partial r^j}{\partial x_m} n_l - \frac{\partial r^j}{\partial x_l} n_m \right) \right] d\Gamma \tag{5.14}$$

A Gaussian quadrature formula can be applied to perform this integration numerically. As in the potential case, care must be taken for the case when any of the sources is close to the element under consideration. In this case more integration points should be used to guarantee accuracy, due to the proximity of the singularity.

Case 2: *Case for which only the source point i is located on the element under consideration.*

This situation involves the computation of an integral containing a logarithmic type of singularity and has already been discussed in chapter 3, for the case of potential problems.

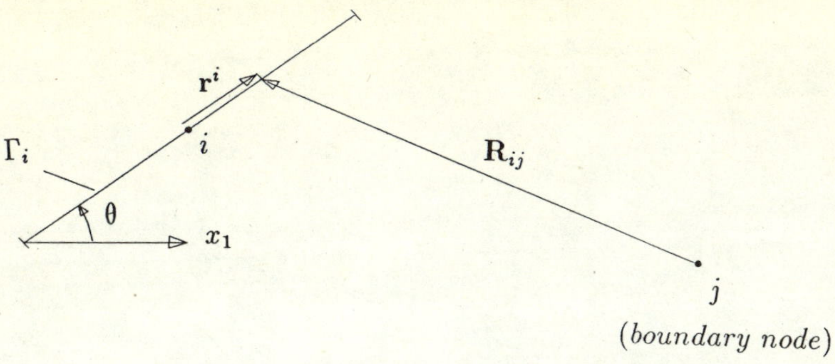

Figure 5.2: Source i located on the element

The integral to be computed is given by equation (5.12) where $\partial r^j / \partial n$ should be taken equal to zero, due to the orthogonality ; r^i and r^j are defined as in figure 5.2.

In order to perform numerically the integrations a change from Cartesian to homogeneous coordinates is necessary. And, as it can be seen from expression (5.12), this case also involves non-singular integrals which can be evaluated without problems using Gaussian quadrature rules.

Case 3: *Case for which only the source point j is located on the element under consideration.*

In this situation, one needs to take into consideration what happens in the limit. To do so, the integration is performed over Γ'_j, and the result is taken to the limit as ε tends to zero. To carry out with this

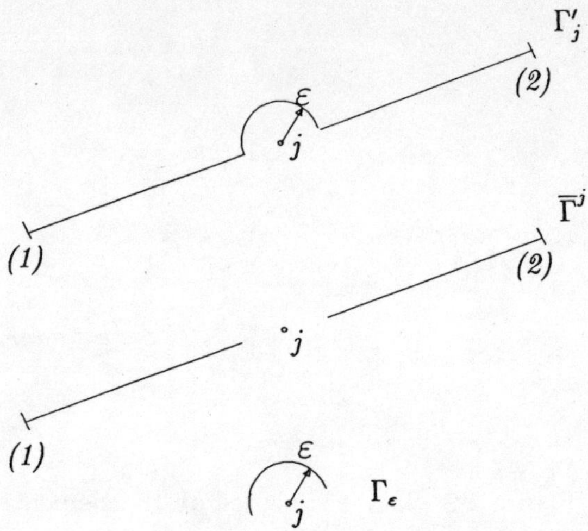

Figure 5.3: Source j located on the element

procedure the boundary Γ'_j is subdivided into two parts such that

$$\Gamma'_j = \overline{\Gamma}^j + \Gamma_\epsilon \tag{5.15}$$

where $\overline{\Gamma}^j$ is the straight part of Γ'_j and Γ_ϵ corresponds to the arc of radius ϵ, centred at j (see figure 5.3).

The integral along Γ'_j can be written as the sum of two integrals as follows:

$$\int_{\Gamma'_j} = \int_{\overline{\Gamma}^j} + \int_{\Gamma_\epsilon} \tag{5.16}$$

Accordingly the integral to be computed, here denoted by $\mathcal{I}^j_{kl}$, is then

$$\mathcal{I}^j_{kl} = \lim_{\epsilon \to 0} \int_{\Gamma'_j} \frac{C_o}{r^j} \left[(3 - 4v) \, \ln(r^i) \, \delta_{km} - \frac{\partial r^i}{\partial x_k} \frac{\partial r^i}{\partial x_m} \right]$$

$$\left\{ \left[(1 - 2v) \, \delta_{lm} + 2 \frac{\partial r^j}{\partial x_l} \frac{\partial r^j}{\partial x_m} \right] \frac{\partial r^j}{\partial n} - (1 - 2v) \left(\frac{\partial r^j}{\partial x_l} n_m - \frac{\partial r^j}{\partial x_m} n_l \right) \right\}$$

$$= \lim_{\epsilon \to 0} \int_{\overline{\Gamma}^j} \frac{C_o}{r^j} \left[(3 - 4v) \, \ln(r^i) \, \delta_{km} - \frac{\partial r^i}{\partial x_k} \frac{\partial r^i}{\partial x_m} \right]$$

$$\left\{\left[(1-2v)\,\delta_{lm}+2\,\frac{\partial r^j}{\partial x_l}\,\frac{\partial r^j}{\partial x_m}\right]\frac{\partial r^j}{\partial n}-(1-2v)\left(\frac{\partial r^j}{\partial x_l}\,n_m-\frac{\partial r^j}{\partial x_m}\,n_l\right)\right\}$$

$$+\lim_{\varepsilon\to 0}\int_{\Gamma_\varepsilon}\frac{C_o}{r^j}\left[(3-4v)\,\ln(r^i)\,\delta_{km}-\frac{\partial r^i}{\partial x_k}\,\frac{\partial r^i}{\partial x_m}\right]$$

$$\left\{\left[(1-2v)\,\delta_{lm}+2\,\frac{\partial r^j}{\partial x_l}\,\frac{\partial r^j}{\partial x_m}\right]\frac{\partial r^j}{\partial n}-(1-2v)\left(\frac{\partial r^j}{\partial x_l}\,n_m-\frac{\partial r^j}{\partial x_m}\,n_l\right)\right\} \tag{5.17}$$

As $\partial r^j/\partial n=0$ on $\overline{\Gamma}^j$, the integral over $\overline{\Gamma}^j$ in equation 5.17, denoted by $\overline{\mathcal{I}}^j_{kl}$, becomes

$$\overline{\mathcal{I}}^j_{kl}=\lim_{\varepsilon\to 0}\int_{\overline{\Gamma}^j}\frac{1-2\nu}{32\pi^2(1-\nu)^2\,E\,r^j}\left[(3-4\nu)\ln(r^i)\,\delta_{km}-\frac{\partial r^i}{\partial x_k}\,\frac{\partial r^i}{\partial x_m}\right]$$

$$\left(\frac{\partial r^j}{\partial x_m}\,n_l-\frac{\partial r^j}{\partial x_l}\,n_m\right) \tag{5.18}$$

This integral corresponds to a Cauchy principal value and can be obtained by applying Kutt's scheme [75, 76, 64] which utilizes finite part integration. Other alternative procedure is to transform the integral in such a way that it is converted into a regular integral plus a simpler Cauchy principal value that can be obtained analytically.

The integral over Γ_ε herein denoted by $\mathcal{I}^\varepsilon_{kl}$ in equation (5.16) represents the jump or discontinuity in $\mathcal{I}^j_{kl}$ when approaching the boundary. It will now be evaluated analytically, in what follows.

From figure 5.4 it can be seen that the following relationships hold:

$$r^j = \varepsilon \tag{5.19}$$

$$\frac{\partial r^j}{\partial n} = \cos(\mathbf{r}^j,\mathbf{n})=-1 \tag{5.20}$$

$$\delta\Gamma = -\varepsilon\,d\theta \tag{5.21}$$

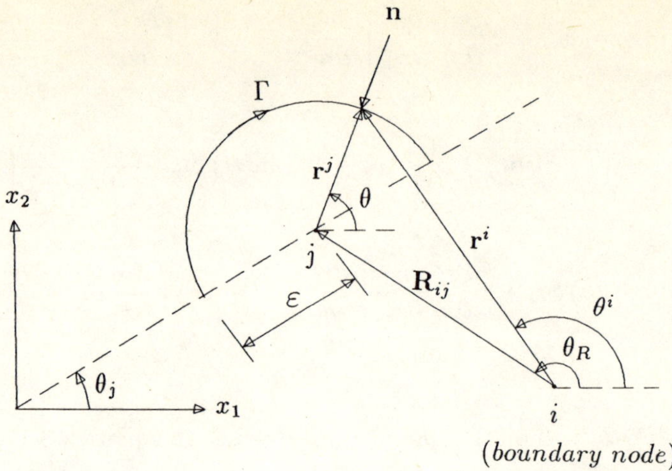

(*boundary node*)

Figure 5.4: The boundary Γ_ε

Substituting these relations into $\mathcal{I}^\varepsilon_{kl}$ yields

$$\mathcal{I}^\varepsilon_{kl} = C_o \lim_{\varepsilon \to 0} \int_{\Gamma_\varepsilon} \left[(3 - 4\nu) \ln r^i \, \delta_{km} - \frac{\partial r^i}{\partial x_k} \frac{\partial r^i}{\partial x_m} \right]$$

$$\left\{ \left[(1 - 2\nu) \, \delta_{lm} + 2 \frac{\partial r^j}{\partial x_l} \frac{\partial r^j}{\partial x_m} \right] + (1 - 2\nu) \left(\frac{\partial r^j}{\partial x_l} n_m - \frac{\partial r^j}{\partial x_m} n_l \right) \right\} d\theta$$

$$(5.22)$$

The integrals $\mathcal{I}^\varepsilon_{kl}$ $(k, l = 1, 2)$ will now be written separately, i.e.

$$\mathcal{I}^\varepsilon_{11} = C_o \lim_{\varepsilon \to 0} \int_{\theta_j + \pi}^{\theta_j} \left\{ \left[(3 - 4\nu) \ln r^i - \left(\frac{\partial r^i}{\partial x_1} \right)^2 \right] \left[(1 - 2\nu) + 2 \left(\frac{\partial r^j}{\partial x_1} \right)^2 \right] \right.$$

$$\left. - \frac{\partial r^i}{\partial x_1} \frac{\partial r^i}{\partial x_2} \left[2 \frac{\partial r^j}{\partial x_1} \frac{\partial r^j}{\partial x_2} + (1 - 2\nu) \left(\frac{\partial r^j}{\partial x_1} n_2 - \frac{\partial r^j}{\partial x_2} n_1 \right) \right] \right\} d\theta$$

$$(5.23)$$

$$\mathcal{I}_{12}^{\varepsilon} = C_o \lim_{\varepsilon \to 0} \int_{\theta_j + \pi}^{\theta_j} \left\{ \left[(3 - 4\nu) \ln r^i - \left(\frac{\partial r^i}{\partial x_1} \right)^2 \right] \right.$$

$$\left[2 \frac{\partial r^j}{\partial x_1} \frac{\partial r^j}{\partial x_2} + (1 - 2\nu) \left(\frac{\partial r^j}{\partial x_2} n_1 - \frac{\partial r^j}{\partial x_1} n_2 \right) \right] -$$

$$\left. \frac{\partial r^i}{\partial x_1} \frac{\partial r^i}{\partial x_2} \left[(1 - 2\nu) + 2 \left(\frac{\partial r^j}{\partial x_2} \right)^2 \right] \right\} d\theta \qquad (5.24)$$

$$\mathcal{I}_{21}^{\varepsilon} = C_o \lim_{\varepsilon \to 0} \int_{\theta_j + \pi}^{\theta_j} \left\{ -\frac{\partial r^i}{\partial x_1} \frac{\partial r^i}{\partial x_2} \left[(1 - 2\nu) + 2 \left(\frac{\partial r^j}{\partial x_1} \right)^2 \right] + \right.$$

$$\left[(3 - 4\nu) \ln r^i - \frac{\partial r^i}{\partial x_1} \frac{\partial r^i}{\partial x_2} \right] \left[2 \frac{\partial r^j}{\partial x_1} \frac{\partial r^j}{\partial x_2} + \right.$$

$$\left. \left. (1 - 2\nu) \left(\frac{\partial r^j}{\partial x_1} n_2 - \frac{\partial r^j}{\partial x_2} n_1 \right) \right] \right\} d\theta \quad (5.25)$$

$$\mathcal{I}_{22}^{\varepsilon} = C_o \lim_{\varepsilon \to 0} \int_{\theta_j + \pi}^{\theta_j} \left\{ -\frac{\partial r^i}{\partial x_1} \frac{\partial r^i}{\partial x_2} \right.$$

$$\left[2 \frac{\partial r^j}{\partial x_1} \frac{\partial r^j}{\partial x_2} + (1 - 2\nu) \left(\frac{\partial r^j}{\partial x_2} n_1 - \frac{\partial r^j}{\partial x_1} n_2 \right) \right] +$$

$$\left. \left[(3 - 4\nu) \ln r^i - \left(\frac{\partial r^i}{\partial x_2} \right)^2 \right] \left[1 - 2\nu + 2 \left(\frac{\partial r^j}{\partial x_2} \right)^2 \right] \right\} d\theta \qquad (5.26)$$

where θ_j is the angle between the element Γ_j and the global coordinate axis x_1.

After taking into account that (figure 5.4)

$$\frac{\partial r^i}{\partial x_1} = \cos \theta^i \qquad (5.27)$$

$$\frac{\partial r^i}{\partial x_2} = \sin\theta^i \tag{5.28}$$

$$\frac{\partial r^j}{\partial x_1} = \cos\theta \tag{5.29}$$

$$\frac{\partial r^j}{\partial x_2} = \sin\theta \tag{5.30}$$

$$n_1 = -\cos\theta \tag{5.31}$$

$$n_2 = \sin\theta \tag{5.32}$$

and with C_o given by (5.10), the integrations can be performed resulting in the following expressions:

$$\mathcal{I}_{11}^\varepsilon = \frac{(1-\nu)(1-2\nu)\,\pi}{16\pi^2(1-\nu)^2\,E}\left[-(3-4\nu)\,\ln R_{ij} + \cos^2\theta_R\right]$$

$$\mathcal{I}_{12}^\varepsilon = \mathcal{I}_{21}^\varepsilon = \frac{(1-\nu)(1-2\nu)\,\pi}{32\pi^2(1-\nu)^2 E}\sin(2\theta_R)$$

$$\mathcal{I}_{22}^\varepsilon = \frac{(1-\nu)(1-2\nu)\,\pi}{16\pi^2(1-\nu)^2 E}\left[-(3-4\nu)\,\ln R_{ij} + \sin^2\theta_R\right] \tag{5.33}$$

In these expressions R_{ij} is the distance between node i to node j and θ_R is the angle between the vector $\mathbf{R}_{ij}$ and the x_1 axis as shown in figure 5.4.

Note: In all previous equations the indices i and j ($i,j = 1,2,\ldots,N$) refer to the source points i and j and do not represent the Cartesian coordinate axis, which are there represented by k, l and m.

Adding the results for $\mathcal{I}_{kl}^\varepsilon$ and $\overline{\mathcal{I}}_{kl}^j$ yields in the limit the final value of the integral defined by (5.16), i.e

$$\mathcal{I}_{kl}^j = \mathcal{I}_{kl}^\varepsilon + \overline{\mathcal{I}}_{kl}^j \tag{5.34}$$

Case 4 : *Case for which both source points i and j coincide and are located on the element under consideration.*

This case occurs in the evaluation of the submatrices $\mathcal{F}^{ij}$ when i is equal to j. See equation (5.4).

Physical considerations can be applied in this case to avoid the treatment of the resulting strongly singular integrals. Rigid body motion considerations will be used here to obtain the submatrices $\mathcal{F}^{ii}$. This technique is widely applied in the conventional boundary element method, to avoid direct computation of some terms [64, 65]. The computation of the main submatrices $\mathcal{F}^{ii}$ through this approach will be explained in section 5.6.

5.2.2 Matrix G for Constant Elements

The matrix **G** in this formulation is exactly equal to the matrix that multiplies the tractions in the classical boundary elements method, also called **G** by most authors [40, 64, 65].

The definition of matrix **G** is given by equation (4.64). Thus, in the case of constant elements, it is possible to write a generic entry of **G**, g_{pq}, as

$$g_{pq} = \lim_{\varepsilon \to 0} \int_{\Gamma'_e} u^{*i}_{kl} \, d\Gamma \qquad p, q = 1, 2, ..., 2N \qquad (5.35)$$

in which

$$\begin{aligned} p &= 2i - 2 + k & k &= 1, 2 \\ q &= 2e - 2 + l & l &= 1, 2 \end{aligned} \qquad (5.36)$$

where i is the boundary node where the source point is located and e is the element where the integration is performed. In the computation of the matrix **G**, two situations should be considered:

i) *Integration over elements which do not include the source point i.*

In this case, the integrals are not singular and can be performed either analytically or numerically. In this work, they have been computed using a Gauss-Legendre quadrature formula.

ii) *Integration over elements which do not include the source point i.*

According to the expression for the fundamental solution, the elements g_{pp} and $g_{p+1\,p+1}$ $\quad(p = 1,3,5,\cdots,2N-1)$ of the matrix **G** contain a logarithmic type of singularity. This is because in these cases, the element where the integral is evaluated includes the source point. Their analytical integration is simple and can be found in many BEM text books [40, 69, 64, 65]. For the sake of completeness they will be shown here together with the analytical integration for $g_{p\,p+1}$ and $g_{p+1\,p}$, which do not include any singularities.

Substituting the fundamental solution into equation (5.35), when $p = q$, gives

$$g_{pp} = \lim_{\varepsilon \to 0} \left\{ \frac{1}{8\pi E(1-\nu)} \left[(3-4\nu) \int_{\Gamma_i} \ln\left(\frac{1}{r^i}\right) d\Gamma + \int_{\Gamma_i} \left(\frac{\partial r^i}{\partial x_1}\right)^2 d\Gamma \right] \right\}$$

$$g_{p\,p+1} = g_{p+1\,p} = \frac{1}{8\pi E(1-\nu)} \int_{\Gamma_i} \frac{\partial r^i}{\partial x_1} \frac{\partial r^i}{\partial x_2} d\Gamma$$

$$g_{p+1\,p+1} = \lim_{\varepsilon \to 0} \left\{ \frac{1}{8\pi E(1-\nu)} \left[(3-4\nu) \int_{\Gamma_i} \ln\left(\frac{1}{r^i}\right) d\Gamma + \int_{\Gamma_i} \left(\frac{\partial r^i}{\partial x_1}\right)^2 d\Gamma \right] \right\}$$

Notice that r^i is the distance between the field point x and the source point, both located on the element Γ_i. The source point coincides with the node i which is the midpoint of the element Γ_i (see figure 5.2).

By geometric considerations (figure 5.2) one has

$$\frac{\partial r^i}{\partial x_1} = \cos\theta_i \tag{5.37}$$

$$\frac{\partial r^i}{\partial x_2} = \sin\theta_i \tag{5.38}$$

Taking limits around the singularities and substituting the relations (5.37) and (5.38) yields

$$g_{pp} = \frac{\ell_i}{8\pi E(1-\nu)}\left[(3-4\nu)(1-\ln\frac{\ell_i}{2})\cos^2\theta_i\right] \tag{5.39}$$

$$g_{p\,p+1} = g_{p+1\,p} = \frac{\ell_i}{8\pi E(1-\nu)}\sin\theta_i\cos\theta_i \tag{5.40}$$

$$g_{p+1\,p+1} = \frac{\ell_i}{8\pi E(1-\nu)}\left[(3-4\nu)\left(1-\ln\frac{\ell_i}{2}\right)+\sin^2\theta_i\right] \tag{5.41}$$

where ℓ_i is the length of the element Γ_i.

5.2.3 Matrix L for Constant Elements

For a constant element the matrix $\mathbf{L}$ is a $2N \times 2N$ diagonal matrix given by

$$\mathbf{L} = \begin{bmatrix} \ell_1 & & & & & & & \\ & \ell_1 & & & & & & \\ & & \ell_2 & & & & & \\ & & & \ell_2 & & & & \\ & & & & \ddots & & & \\ & & & & & \ddots & & \\ & & & & & & \ell_N & \\ & & & & & & & \ell_N \end{bmatrix} \tag{5.42}$$

where ℓ_k $(k=1,2,....,N)$ are the element lengths.

No integrations have to be carried out in this case (i.e. for constant elements) and the operations involving this matrix are very simple.

5.2.4   Load Vector for Constant Elements

The load vector $\mathbf{Q}$ is defined by equation (4.79), i.e.

$$\mathbf{Q} = \overline{\mathbf{P}} + \mathbf{R}^T \, \mathbf{B} \tag{5.43}$$

where $\overline{\mathbf{P}}$, $\mathbf{R}$ and $\mathbf{B}$ are given by formulae (4.66), (4.78) and (4.67).

It will be assumed here that the prescribed tractions over the element Γ_k are constant. The vector of prescribed loads on the element $\overline{\mathbf{Q}}^k$ can then be directly obtained by the following equation

$$\overline{\mathbf{Q}}^k = \left\{ \begin{array}{c} \overline{Q}_1^k \\ \overline{Q}_2^k \end{array} \right\} = \left\{ \begin{array}{c} \overline{p}_1^k \, \ell_k \\ \overline{p}_2^k \, \ell_k \end{array} \right\} \tag{5.44}$$

where $\overline{p}_1^k$ and $\overline{p}_2^k$ are the average or constant prescribed tractions in the directions of the coordinates on the element Γ_k.

For problem with no body forces, the vector $\mathbf{B}$ is the null vector and hence the vector of equivalent nodal loads $\mathbf{Q}$ is equal to $\overline{\mathbf{Q}}$.

The evaluation of the vector $\mathbf{B}$ in the case of non-zero body forces will be treated in section 5.5.

5.3   The Quadratic Element

In engineering practice, it is important to solve problems that involve the bending of elastic bodies. In such cases, and in other elastic problems, the displacement and tractions can vary rapidly along the boundary. The approximation of those variables by quadratic or higher order functions is, therefore, important to assure the accuracy of the solution.

In this section, the computer implementation of the hybrid displacement boundary element formulation will be described for the case of isoparametric quadratic elements.

The utilization of quadratic elements is also important when dealing with curved boundaries, as in such cases it is not possible to have a good approximation for the boundary, with lower order elements.

The standard quadratic finite element type interpolation functions are used here to interpolate the displacements and tractions $-\tilde{u}$ and $\tilde{p}-$ over the element, in terms of nodal values. Hence 3 nodes per element must be defined and these are chosen to be the two end points and the midpoint of the element.

Accordingly, the displacement vector over the element is approximated by

$$\tilde{u} = \left\{ \begin{array}{c} \tilde{u}_1 \\ \tilde{u}_2 \end{array} \right\} = \left[\begin{array}{cccccc} \varphi_1 & 0 & \varphi_2 & 0 & \varphi_3 & 0 \\ 0 & \varphi_1 & 0 & \varphi_2 & 0 & \varphi_3 \end{array} \right] \left\{ \begin{array}{c} u_1^1 \\ u_2^1 \\ u_1^2 \\ u_2^2 \\ u_1^3 \\ u_2^3 \end{array} \right\} \tag{5.45}$$

where u_i^k is the i component of the displacement vector at the node k on the element, $(k = 1, 2, 3)$. Notice that, for simplicity, u_i^k is written without the tilde, which denotes boundary. The interpolation functions φ_i have been defined in terms of the homogeneous coordinate η in chapter 3. They are repeated here for completeness:

$$\varphi_1 = \frac{1}{2}\eta(\eta - 1)$$

$$\varphi_2 = \frac{1}{2}(1 - \eta^2)$$

$$\varphi_3 = \frac{1}{2}\eta(\eta + 1) \tag{5.46}$$

The element is isoparametric and, as such, its geometry is also approximated by the same interpolation functions φ_1, φ_2 and φ_3. The global coordinates of a point on the element, x_1 and x_2, can be expressed as:

$$\left\{ \begin{array}{c} x_1 \\ x_2 \end{array} \right\} = \left[\begin{array}{cccccc} \varphi_1 & 0 & \varphi_2 & 0 & \varphi_3 & 0 \\ 0 & \varphi_1 & 0 & \varphi_2 & 0 & \varphi_3 \end{array} \right] \left\{ \begin{array}{c} x_1^1 \\ x_2^1 \\ x_1^2 \\ x_2^2 \\ x_1^3 \\ x_2^3 \end{array} \right\} \tag{5.47}$$

where x_i^k is the coordinate x_i of node k on the element (k=1,2,3).

When performing the boundary integrals, there is the need for a change of variables because the integral is defined along the boundary Γ and the shape functions are given in terms of η. The Jacobian of the transformation of variables, from Γ to η, which was given in section 3.3, is:

$$J = \frac{d\Gamma}{d\eta} \tag{5.48}$$

and can be expressed in terms of nodal Cartesian coordinates as:

$$J = \left\{ [(x_1^1 - 2x_1^2 + x_1^3)\eta + \frac{1}{2}(x_1^3 - x_1^1)]^2 + [(x_2^1 - 2x_2^2 + x_2^3)\eta + \frac{1}{2}(x_2^3 - x_2^1)]^2 \right\}^{\frac{1}{2}} \tag{5.49}$$

Now the evaluation of the matrices $\mathbf{F}$, $\mathbf{G}$, $\mathbf{L}$ and the vectors $\overline{\mathbf{P}}$ and $\mathbf{B}$, which are necessary for the numerical solution of the problem, will be discussed.

5.3.1 Matrix F for Quadratic Elements

The matrix $\mathbf{F}$, which is used to compute the boundary stiffness matrix $\mathbf{K}$, does not depend on the type of approximation used for the boundary variables $\tilde{u}$ and $\tilde{p}$. It varies with the geometry of the element and with the type of fundamental solution adopted. In this work, the Kelvin fundamental solution is used for both the constant and the quadratic element. Therefore the differences in the matrix $\mathbf{F}$, between the two cases, is only due to the differences in the geometries of the element.

The considerations in section 5.2.1 for the constant element, up to equation (5.9), are valid for the quadratic element as well. Thus a generic coefficient f_{kl}^{ij} of the submatrices $\mathcal{F}^{ij}$ of $\mathbf{F}$ is given by

$$f_{kl}^{ij} = \sum_{e=1}^{Ne} \lim_{\varepsilon \to 0} \int_{\Gamma_e'} \frac{C_o}{r^j} \left[(3 - 4v) \ln r^i \delta_{km} - \frac{\partial r^i}{\partial x_k} \frac{\partial r^i}{\partial x_m} \right]$$

$$\left\{ \left[(1 - 2v)\delta_{lm} + 2\frac{\partial r^j}{\partial x_l} \frac{\partial r^j}{\partial x_m} \right] \frac{\partial r^j}{\partial n} - (1 - 2v) \left(\frac{\partial r^j}{\partial x_l} n_m - \frac{\partial r^j}{\partial x_m} n_l \right) \right\} \quad (5.50)$$

in which Ne is the number of boundary elements; the boundary Γ_e' is the boundary corresponding to the element Γ_e in the domain Ω' The subscripts i and j mean that the unit loads are applied at the boundary nodes i and j, i.e. i and j correspond to source points.

The location of these source points in relation to the element over which the integration is being performed, defines the type of integral to be treated. In fact, it can be seen from equation (5.9) that the source point i produces a logarithmic type of singularity when the source point i and the field point x over the element coincide, namely when $r^i = 0$. There is another singularity, of the type $1/|r^j|$, due to the source point at the node j. A third kind of singular integral also occurs when both source points i and j coincide, and this is stronger than the other previous two singularities.

These four different types of integrals, which happen depending on the position of the source points, will now be treated in detail. Five distinct cases should be considered, as follows:

Case 1: *Integration over the element Γ_e which includes neither source point i nor source point j*

All the functions to be integrated are well defined, thus no singularities occur in this situation. There is no need for any limit consideration and the integral to be evaluated can then be written as

$$\int_{\Gamma_e} \frac{C_o}{r^j} \left[(3 - 4v) \ln r^i \delta_{km} - \frac{\partial r^i}{\partial x_k} \frac{\partial r^i}{\partial x_m} \right]$$

$$\left\{ \left[(1 - 2v)\delta_{lm} + 2\frac{\partial r^j}{\partial x_l} \frac{\partial r^j}{\partial x_m} \right] \frac{\partial r^j}{\partial n} - (1 - 2v) \left(\frac{\partial r^j}{\partial x_l} n_m - \frac{\partial r^j}{\partial x_m} n_l \right) \right\}$$

$$(5.51)$$

where $i, j \notin \Gamma_k$, and the variables involved are depicted in figure 5.5.

A standard Gauss-Legendre quadrature rule giving accurate results can be applied in this case. When any of the sources i and j, or both of them, are close to Γ_k the integrand varies rapidly with the distances from the sources to the element. The number of integration points should, therefore, be chosen carefully by testing the convergence of different numerical integrations. After carrying out a series

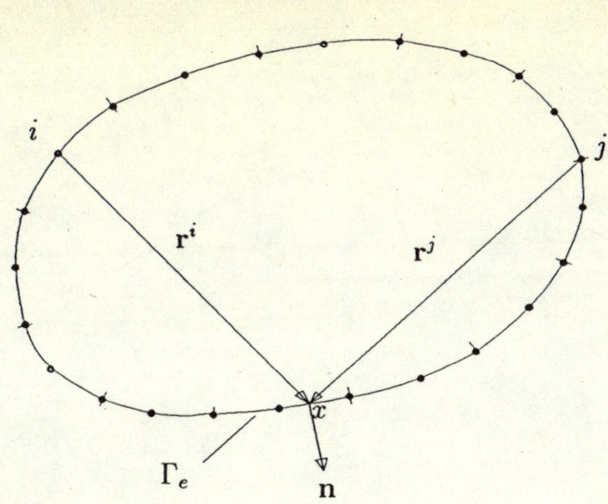

Figure 5.5: Both sources i and j outside the element

of tests, it was found that twelve integration points were sufficient, if the distances between the source points and the nearest point on the element is greater than the distance from the mid node to the end point closer to it.

Case 2: *Integration over the element Γ_e which includes only the source point i.*

In this case, the integral involves a logarithmic singularity which is weak and hence the integral is well defined. The integral to be computed in this case is expressed as

$$\int_{\Gamma_e} \frac{C_o}{r^j} \left[(3 - 4v) \ln r^i \delta_{km} - \frac{\partial r^i}{\partial x_k} \frac{\partial r^i}{\partial x_m} \right]$$

$$\left\{ \left[(1 - 2v)\delta_{lm} + 2\frac{\partial r^j}{\partial x_l} \frac{\partial r^j}{\partial x_m} \right] \frac{\partial r^j}{\partial n} - (1 - 2v) \left(\frac{\partial r^j}{\partial x_l} n_m - \frac{\partial r^j}{\partial x_m} n_l \right) \right\}$$

$$(5.52)$$

where $i \in \Gamma_e$ and the variables involved are shown in figure 5.6, for this particular situation.

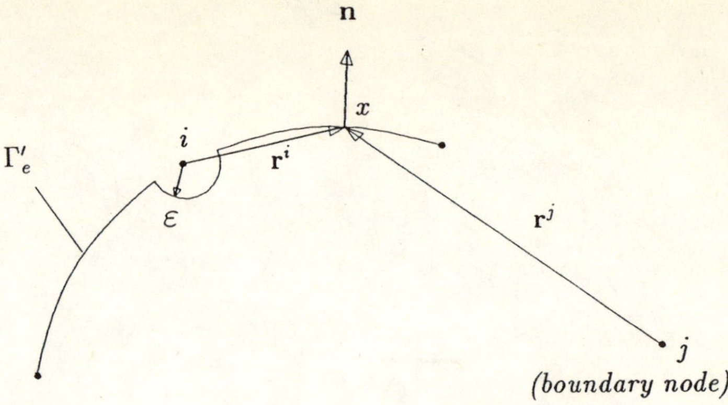

Figure 5.6: The element includes the source point i

The logarithmic terms can be integrated by an appropriate scheme such as a logarithmic Gaussian quadrature formula [67] or the scheme proposed by Telles [68]. The other nonsingular terms can be integrated as in case 1.

Case 3: *Integration over the element Γ_e which includes only the source point j.*

Now limit considerations have to be taken into account, because the integrand contains singular terms when ε tends to zero. The integral whose value has to be computed is

$$\lim_{\varepsilon \to 0} \int_{\Gamma'_e} \frac{C_o}{r^j} \left[(3 - 4v) \ln r^i \delta_{km} - \frac{\partial r^i}{\partial x_k} \frac{\partial r^i}{\partial x_m} \right]$$

$$\left\{ \left[(1 - 2v)\delta_{lm} + 2\frac{\partial r^j}{\partial x_l} \frac{\partial r^j}{\partial x_m} \right] \frac{\partial r^j}{\partial n} - (1 - 2v) \left(\frac{\partial r^j}{\partial x_l} n_m - \frac{\partial r^j}{\partial x_m} n_l \right) \right\}$$

$$(5.53)$$

In order to perform the integration, the boundary Γ'_e is subdivided into two parts (figure 5.7):

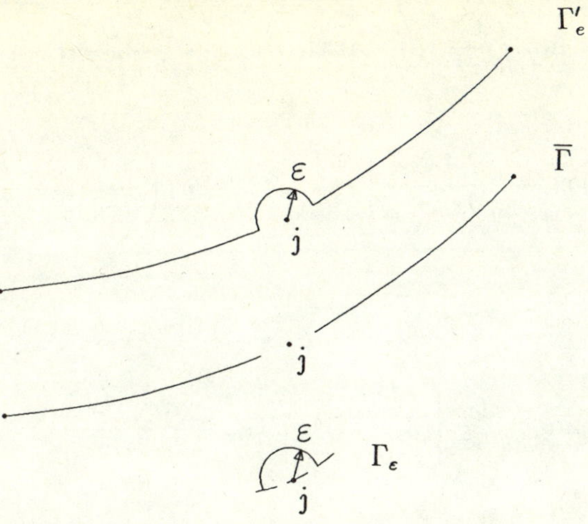

Figure 5.7: The element includes the source point j

— The first one is the boundary corresponding to Γ_e after excluding the singularity; here it is denoted by $\bar{\Gamma}$.

— The second part is the arc Γ_ε, as depicted in figure 5.7 where the source point j is shown to coincide with the midpoint of the element. This point j may as well be located at the end nodes.

It is then possible to write

$$\mathcal{J}_{kl} = \overline{\mathcal{J}}_{kl} + \mathcal{J}_{kl}^{\varepsilon} \tag{5.54}$$

where

$$\mathcal{J}_{kl} = \lim_{\varepsilon \to 0} \int_{\Gamma_e'} \frac{C_o}{r^j} \left[(3 - 4v) \ln r^i \delta_{km} - \frac{\partial r^i}{\partial x_k} \frac{\partial r^i}{\partial x_m} \right]$$

$$\left\{ \left[(1 - 2v)\delta_{lm} + 2\frac{\partial r^j}{\partial x_l} \frac{\partial r^j}{\partial x_m} \right] \frac{\partial r^j}{\partial n} - (1 - 2v) \left(\frac{\partial r^j}{\partial x_l} n_m - \frac{\partial r^j}{\partial x_m} n_l \right) \right\} \tag{5.55}$$

132

$$\overline{\mathcal{J}}_{kl} = \lim_{\varepsilon \to 0} \int_{\Gamma} \frac{C_o}{r^j} \left[(3 - 4v) \ln r^i \delta_{km} - \frac{\partial r^i}{\partial x_k} \frac{\partial r^i}{\partial x_m} \right]$$

$$\left\{ \left[(1 - 2v)\delta_{lm} + 2\frac{\partial r^j}{\partial x_l} \frac{\partial r^j}{\partial x_m} \right] \frac{\partial r^j}{\partial n} - (1 - 2v) \left(\frac{\partial r^j}{\partial x_l} n_m - \frac{\partial r^j}{\partial x_m} n_l \right) \right\}$$

(5.56)

and

$$\mathcal{J}_{kl}^{\varepsilon} = \lim_{\varepsilon \to 0} \int_{\Gamma'_e} \frac{C_o}{r^j} \left[(3 - 4v) \ln r^i \delta_{km} - \frac{\partial r^i}{\partial x_k} \frac{\partial r^i}{\partial x_m} \right]$$

$$\left\{ \left[(1 - 2v)\delta_{lm} + 2\frac{\partial r^j}{\partial x_l} \frac{\partial r^j}{\partial x_m} \right] \frac{\partial r^j}{\partial n} - (1 - 2v) \left(\frac{\partial r^j}{\partial x_l} n_m - \frac{\partial r^j}{\partial x_m} n_l \right) \right\}$$

(5.57)

Integration over $\overline{\Gamma}$

In expression (5.56) the only singular term is the one that involves

$$\frac{1}{r^j} \left(\frac{\partial r^j}{\partial x_l} n_m - \frac{\partial r^j}{\partial x_m} n_l \right)$$

(5.58)

and whose integral has a Cauchy principal value, since the expression inside the parenthesis changes sign across the point j. When the source point j is located at an extreme node of the element, one must match two adjacent elements to obtain the principal value. The integration of all the other terms can be handled accurately by a Gaussian quadrature scheme, as in case 1.

Integration over Γ_ε

According to the position of the source point j on the element, two different cases can occur:

i) j is at one of the end nodes of the element.

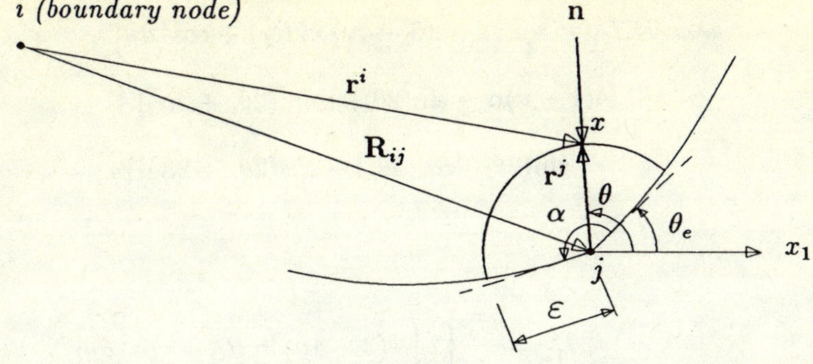

Figure 5.8: Singularity j at an end node

From figure 5.8 one can establish the following relations:

$$r^j = \epsilon \tag{5.59}$$

$$n_1 = -\cos\theta \tag{5.60}$$

$$n_2 = -\sin\theta \tag{5.61}$$

$$\frac{\partial r^j}{\partial x_1} = \cos\theta \tag{5.62}$$

$$\frac{\partial r^j}{\partial x_2} = \sin\theta \tag{5.63}$$

$$\frac{\partial r^i}{\partial x_1} = \cos\theta^i \tag{5.64}$$

$$\frac{\partial r^i}{\partial x_2} = \sin\theta^i \tag{5.65}$$

$$\frac{\partial r^j}{\partial n} = -1 \tag{5.66}$$

$$d\Gamma = -\epsilon\, d\theta \tag{5.67}$$

These relations should then be inserted into formula (5.57), which after being integrated and taken to the limit yields the final expression for the computation of $\mathcal{J}_{kl}^{\varepsilon}$, as shown below:

$$\mathcal{J}_{11}^{\varepsilon} = \frac{C_o}{4}\left\{2\left[-(3-4\nu)\ln R_{IJ} + \cos^2\theta_R\right]\right.$$

$$[4(1-\nu)\alpha - \sin(2\theta_e) + \sin(2\theta_e + 2\alpha)] +$$

$$\left.\sin(2\theta_R)[\cos(2\theta_e) - \cos(2\theta_e + 2\alpha)]\right\}. \qquad (5.68)$$

$$\mathcal{J}_{12}^{\varepsilon} = \frac{C_o}{4}\left\{2\left[-(3-4\nu)\ln R_{ij} + \cos^2\theta_R\right]\right.$$

$$[\cos(2\theta_e) - \cos(2\theta_e + 2\alpha)] +$$

$$\left.\sin(2\theta_R)\left[4(1-\nu)\alpha + \sin(2\theta_e) - \sin(2\theta_e + 2\alpha)\right]\right\} \qquad (5.69)$$

$$\mathcal{J}_{21}^{\varepsilon} = \frac{C_o}{4}\left\{2\left[-(3-4\nu)\ln R_{ij} + \sin^2\theta_R\right]\right.$$

$$[\cos(2\theta_e) - \cos(2\theta_e + 2\alpha)] +$$

$$\left.2\sin(2\theta_R)\left[2(1-\nu)\alpha - \sin(2\theta_e) + \sin(2\theta_e + 2\alpha)\right]\right\} \qquad (5.70)$$

$$\mathcal{J}_{22}^{\varepsilon} = \frac{C_o}{4}\left\{2\left[-(3-4\nu)\ln R_{ij} + \sin^2\theta_R\right]\right.$$

$$[4(1-\nu)\alpha + \sin(2\theta_e) - \sin(2\theta_e + 2\alpha)] +$$

$$\left.\sin(2\theta_R)[\cos(2\theta_e) - \cos(2\theta_e + 2\alpha)]\right\}; \qquad (5.71)$$

where R_{ij} is the length of the vector $\mathbf{R}$ that goes from the source point i to the source point j; θ_R is the angle between $\mathbf{R}_{ij}$ and the x_1 direction; θ_e is the angle between the tangent to the element whose node 1 coincides with j; α is the angle between the two elements at j (see figure 5.8) and

$$C_o = \frac{1}{32\pi^2(1-\nu)^2\mu}$$

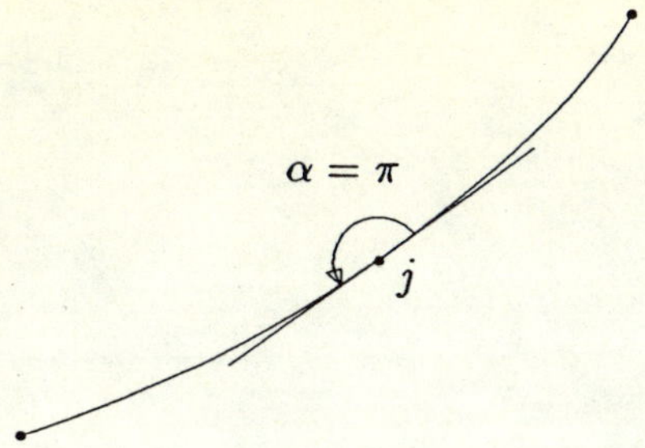

Figure 5.9: Singularity j at the mid node

ii) j is at the mid node of the element.

The expression for the integral $\mathcal{J}_{kl}^{\varepsilon}$ can be found by replacing the angle α by π into the corresponding formulae for case **i**. See figure 5.9. The integrals $\mathcal{J}_{kl}^{\varepsilon}$ $(k,l = 1,2)$ can then be expressed by:

$$\mathcal{J}_{11}^{\varepsilon} = \frac{(1-\nu)(1-2\nu)\,\pi}{16\pi^2(1-\nu)^2\,\mu}\left[-(3-4\nu)\,\ln R_{ij} + \cos^2\theta_R\right] \qquad (5.72)$$

$$\mathcal{J}_{12}^{\varepsilon} = \mathcal{J}_{21}^{\varepsilon} = \frac{(1-\nu)(1-2\nu)\,\pi}{32\pi^2(1-\nu)^2\mu}\,\sin(2\theta_R) \qquad (5.73)$$

$$\mathcal{J}_{22}^{\varepsilon} = \frac{(1-\nu)(1-2\nu)\,\pi}{16\pi^2(1-\nu)^2\mu}\left[-(3-4\nu)\,\ln R_{ij} + \sin^2\theta_R\right] \qquad (5.74)$$

which are the same expressions found for the constant element.

Case 4: *Integration over the element Γ_e including two source points, i and j, which are non-coincident.*

According to the relative position of the source points on the element, six different possibilities should be considered in this case, as shown in figure 5.10.

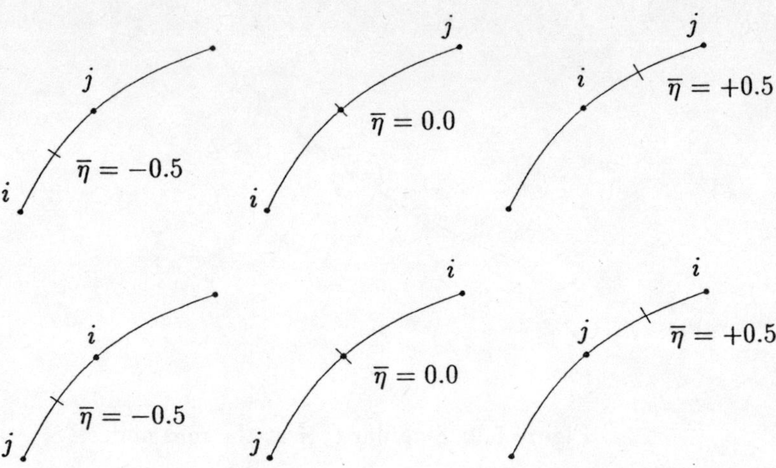

Figure 5.10: Both singularities i and j occurring on the same element

In all these cases two different types of singularities occur on the same element. One singularity is due to source i, which is of a logarithmic type; the other, which is due to source j, is similar to the singularity studied in case 3.

It is possible to avoid the integration of two simultaneous singularities by subdividing the element into two parts, as it has been done in chapter 3 for potential problems. To choose the point where the element should be split apart, namely the value of $\overline{\eta}$ (see figure 5.10), the accuracy and convergence of the singular integrations have to be tested because of the proximity of the other singularity.

In the part of the element which includes the source point i, the integration should be performed as in case 2. The situation in which the source j is included should be handled as in case 3.

Case 5: *Integration over an element having two source points i and j, which are coincident.*

This case occurs in the computation of the main submatrices $\mathcal{F}^{ii}$ of the matrix $\mathbf{F}$.

The rigid body motion technique, which will be explained in section 5.6, can be applied to evaluate $\mathcal{F}^{ii}$, therefore avoiding their direct computation. This physical consideration can be used in exactly the same way for the two types of element studied in this chapter, namely the constant and the quadratic elements.

In this hybrid-displacement approach, the evaluation of the internal displacements and stresses when body forces are present does not require the computation of any volume integral. Internal displacements and stresses are then computed by applying expressions (4.46) and (4.58), respectively, in the same way they would have been evaluated if there were no body forces.

5.3.2 Matrix G for Quadratic Elements

Notice that there is a 2×6 submatrix $\mathcal{G}^e$ in matrix $\mathbf{G}$ corresponding to each position of the source point i on the boundary and to each element Γ_e. This is given by

$$\mathcal{G}^e = \int_{\Gamma_e} \begin{bmatrix} u_{11}^{*i} & u_{12}^{*i} \\ u_{11}^{*i} & u_{12}^{*i} \end{bmatrix} \begin{bmatrix} \phi_1 & 0 & \phi_2 & 0 & \phi_3 & 0 \\ 0 & \phi_1 & 0 & \phi_2 & 0 & \phi_3 \end{bmatrix} d\Gamma \qquad (5.75)$$

The integrals to be evaluated have the form:

$$\int_{\Gamma_e} \phi_k \, u_{lm}^{*i} \, d\Gamma \qquad (5.76)$$

Due to the singularity in the fundamental solution, when the element includes the source point i, two distinct procedures should be followed in the computation of G, as it will now be described.

i) *Integration over an element which does not include the source point i.*

All the integrals to be computed in this case are regular. Their evaluation is then a simple task which is achieved by adopting a proper numerical integration scheme. A Gauss-Legendre quadrature formula can be applied successfully.

When selecting the number of integration points, it is important to consider the influence of the proximity of a singularity. This situation occurs when the source point i is close to the element where the integration is being performed.

ii) *Integration over an element which includes the source point i.*

The evaluation of the submatrices G now involves non-singular and also singular integrals. The singularity is of the logarithmic type; as such, the numerical computation of the singular integrals can be performed by applying a suitable scheme [67, 68]. The non-singular integrals may be computed by a Gaussian quadrature rule without any special consideration.

These submatrices are then conveniently assembled to generate the whole matrix G.

5.3.3   Matrix L for Quadratic Elements

Following the definition of the matrix **L** given by formula 4.65, in the case of quadratic elements, to each element Γ_e corresponds a symmetric 6×6 submatrix $\mathcal{L}^e$ defined as

$$
\mathcal{L}^e = \int_{\Gamma_e}
\begin{bmatrix}
\phi_1^2 & 0 & \phi_1\phi_2 & 0 & \phi_1\phi_3 & 0 \\
 & \phi_1^2 & 0 & \phi_1\phi_2 & 0 & \phi_1\phi_3 \\
 & & \phi_2^2 & 0 & \phi_2\phi_3 & 0 \\
 & & & \phi_2^2 & 0 & \phi_2\phi_3 \\
 & & & & \phi_3^2 & 0 \\
 & & & & & \phi_3^2
\end{bmatrix}
d\Gamma
\qquad (5.77)
$$

The integrals to be computed are all regular and can be easily obtained.

Matrix **L** is then found by properly assembling these submatrices, taking into account the connectivities of the elements.

5.3.4   Load Vector for Quadratic Elements

Formula (4.79) states that the load vector **Q** is given as a summation of two parcels: one due to the prescribed tractions and the other related to body forces, i.e.

$$
\mathbf{Q} = \overline{\mathbf{P}} + \mathbf{R}^T \mathbf{B}
\qquad (5.78)
$$

where $\overline{\mathbf{P}}$ is due to the prescribed tractions and is defined by equation 4.66. The last term on the right-hand side represents the contribution of the body forces.

According to equation (4.78), vector **R** is given by

$$
\mathbf{R} = (\mathbf{G}^T)^{-1} \mathbf{L}
\qquad (5.79)
$$

In section 5.5 procedures to evaluate the vector $\mathbf{B}$ will be described.

The computation of the vector $\overline{\mathbf{P}}$ will now be presented.

The boundary tractions are assumed to vary quadratically. It will be supposed that the prescribed tractions can also be represented through the quadratic interpolation functions ϕ_i, which are used to approximate the boundary displacements and traction, i.e.

$$\overline{\mathbf{p}} = \left\{ \begin{array}{c} \overline{p}_1 \\ \overline{p}_2 \end{array} \right\} = \left[\begin{array}{cccccc} \phi_1 & 0 & \phi_2 & 0 & \phi_3 & 0 \\ 0 & \phi_1 & 0 & \phi_2 & 0 & \phi_3 \end{array} \right] \left\{ \begin{array}{c} \overline{p}_1^1 \\ \overline{p}_2^1 \\ \overline{p}_1^2 \\ \overline{p}_2^2 \\ \overline{p}_1^3 \\ \overline{p}_2^3 \end{array} \right\}^e = \phi^T \, \overline{\mathbf{P}}_e \qquad (5.80)$$

in which $\overline{\mathbf{p}}$ is the vector of prescribed tractions at a point on the boundary; $\overline{p}_l^k$ denotes the l component of the prescribed traction at the node k $(k = 1, 2, 3)$ on the element.

Equation (5.80) is now taken into consideration to define vector $\overline{\mathbf{P}}$, in expression (4.66). The contribution of an element Γ_e to the vector of equivalent prescribed tractions, $\overline{\mathbf{P}}$, can therefore be evaluated as

$$\left\{ \begin{array}{c} \overline{\mathbf{P}}_e^1 \\ \overline{\mathbf{P}}_e^2 \\ \overline{\mathbf{P}}_e^3 \end{array} \right\} = \left(\int_{\Gamma_e} \phi \, \phi^T \, d\Gamma \right) \overline{\mathbf{P}}_e \qquad (5.81)$$

where $\overline{\mathbf{P}}_e^i$ $(i = 1, 2, 3)$ represents the contribution of the element Γ_e to the vector of equivalent prescribed tractions at the node i on the element.

The contribution of two adjacent elements should be added when assembling the contribution of all the elements to generate $\overline{\mathbf{P}}$.

Observe that the integrals in (5.81) have already been computed when evaluating matrix **L**.

5.4   Computation of the Submatrices $\mathcal{F}^{ii}$

The coefficients f_{kl}^{ii} of the main submatrices $\mathcal{F}^{ii}$ corresponds to the main diagonal and its neighbour terms in matrix **F**, according to the following identities:

$$f_{11}^{ii} = F_{pp} \tag{5.82}$$

$$f_{12}^{ii} = F_{pp+1} \tag{5.83}$$

$$f_{21}^{ii} = F_{p+1p} \tag{5.84}$$

$$f_{22}^{ii} = F_{p+1p+1} \tag{5.85}$$

where $p = 2i - 1$ and i is the number of the boundary node where the unit source is located. Therefore as $i = 1, 2, 3, \ldots, N, \quad p = 1, 3, 5, \ldots, 2N - 1$.

By assuming a rigid body translation in the direction of one of the Cartesian coordinate axis, the tractions and body forces are zero and, therefore, the vector **Q** is the null vector. Equation (4.76) then becomes

$$\mathbf{K} \mathbf{I}^q = 0 \tag{5.86}$$

Where $\mathbf{I}^q$ is a vector that for all nodes has unit displacement along the q direction ($q = 1, 2$) and zero displacement in any other direction.

Taking into account that

$$\mathbf{K} = \mathbf{R}^T \mathbf{F} \mathbf{R}$$

it is possible to write

$$\mathbf{R}^T \mathbf{F} \mathbf{R} \mathbf{I}^q = 0 \tag{5.87}$$

and also, since $\mathbf{R}^T$ is non-singular,

$$\mathbf{F}\,\mathbf{R}\,\mathbf{I}^q = 0 \qquad\qquad (5.88)$$

If two different translations are supposed to occur, in the direction of the two coordinate axis, expression (5.88) produces $2N$ algebraic equations. These equations when solved give the $2N$ unknown coefficients, $F_{pp}\ F_{p\,p+1}\ F_{p+1\,p}\ F_{p+1\,p+1}\quad (p = 1, 3, 5, \ldots, 2N - 1)$, which correspond to the submatrices $\mathcal{F}^{ii}$.

Note:

In chapter 4 it has been proved that the matrix $\mathbf{F}$ is symmetric. If the $\mathcal{F}^{ii}$ coefficients were evaluated by the integrals that define the elements of the matrix $\mathbf{F}$, they would not have finite values. Since this is the result of the hypothesis of concentrated fictitious loads (which is a mathematical idealization), it seems reasonable to apply a physical consideration for their evaluation. The consideration of rigid body motion produces finite values for these coefficients. This reasoning is similar to the one that considers only the "finite part" of hyper singular coefficients in some formulations [92].

When rigid body considerations are implemented in computer code, non-symmetric submatrices $\mathcal{F}^{ii}$ will appear in general. This fact would destroy the symmetry of the stiffness matrix, which is a useful characteristic of this hybrid formulation. Fortunately, however, the symmetry of the matrix $\mathbf{F}$ can be imposed together with the rigid body translation properties, by a least squares procedure. A symmetric stiffness matrix is then obtained and the formulation can represent a rigid body movement in a least square sense. It is important to point out here that the solution is very stable and does not appreciably change by imposing the symmetry conditions in the coefficients of the submatrices in the diagonal of the matrix $\mathbf{F}$. It is important to point out that all the coefficients of matrix $\mathbf{F}$ which are computed via integrations are symmetric, as expected.

In order to check whether the model represents properly rigid body movements after the imposition of the symmetry condition, the eigenvalues and the eigenvectors of the stiffness matrix have been investigated for many different cases. All the results have shown that the model can properly represent those movements. Even the rigid body rotation mode, which is not explicitly imposed, is satisfied.

5.5   Body Forces

In practice many engineering problems present non-zero body forces. The most commonly encountered are gravitational, thermal and centrifugal loads. For those cases, the vector $\mathbf{B}$ which defined in equation (4.67) should be computed.

Each component B_m of $\mathbf{B}$ corresponds to the application of a unit load at the node i in the direction k, with $m = 2i-2+k$. This unit load produces the fundamental displacement u_{kl}^{*i}. Consequently the B_m component is given by the volume integral shown below:

$$B_m = \int_\Omega b_l \, u_{kl}^{*i} \, d\Omega \qquad (5.89)$$

It can be recognized by the definition of $\mathbf{B}$ that this vector also appears in the classical direct boundary element method formulation [40, 64, 65].

There, as in this hybrid formulation, it corresponds to the domain integrals related to body forces.

The direct computation of these domain integrals can be done numerically by dividing the domain into cells as shown in chapter 3, for potential problems. The internal cell integrals can be satisfactorily computed by a numerical integration formula or, in some simple cases, analytically. An interesting alternative is to employ the semi-analytical integration scheme which has been proposed by Telles and Brebbia in reference [80].

The transformation of the domain integral into a boundary integral is possible when the body force is derivable from a scalar potential whose Laplacian is at most a constant. Fortunately such is the case of the body forces most commonly encountered in practice: constant gravitational loads. The cases of centrifugal forces due to a fixed axis of rotation and of an elastic body under a steady state temperature potential can also be handled in a similar manner. This procedure has been applied by several boundary element researchers [35, 78, 79].

Whenever possible, the direct evaluation of the domain integrals should be avoided. Since it requires the subdivision of the domain into internal cells, a large effort in data preparation is needed. Furthermore the computation of the domain integrals itself consumes a considerable amount of computer time.

5.5.1   Transformation of the Domain Integrals into Boundary Integrals

The approach followed in this section uses the Galerkin tensor, corresponding to the Kelvin fundamental solution, to take the domain integrals to the boundary. This unified procedure can be applied in those cases where the body force is derivable from a scalar potential whose Laplacian is at most a constant. These are the cases of constant gravitational loads, thermal loads due to a steady state temperature field and centrifugal loads corresponding to the rotation about a fixed axis.

The integral to be converted into the boundary is

$$\int_\Omega b_l \, u_{kl}^{*i} \, d\Omega \tag{5.90}$$

Suppose that the body force components b_l can be obtained from a potential function ψ such that

$$b_l = \frac{\partial \psi}{\partial x_l} \tag{5.91}$$

where the potential function ψ has a constant Laplacian, namely

$$\nabla^2 \psi = K_0 = constant \tag{5.92}$$

Substituting (5.91) into the domain integral (5.90) gives

$$B_i = \int_\Omega \frac{\partial \psi}{\partial x_l} \, u_{kl}^{*i} \, d\Omega \tag{5.93}$$

which can be written as

$$B_i = \int_\Omega \frac{\partial}{\partial x_l}(\psi \, u_{kl}^{*i}) \, d\Omega - \int_\Omega \psi \, \frac{\partial u_{kl}^{*i}}{\partial x_l} \, d\Omega \tag{5.94}$$

By applying the divergence theorem in the first integral, it is possible to write

$$B_i = \int_\Gamma \psi \, u^{*i}_{kl} \, n_l \, d\Gamma - \int_\Omega \psi \, \frac{\partial u^{*i}_{kl}}{\partial x_l} \, d\Omega \qquad (5.95)$$

where n_l is the direction cosine of the normal $\mathbf{n}$ to the boundary Γ with respect to the x_l axis.

The domain integral in (5.95) can also be taken to the boundary by using the Galerkin tensor G_{kl} to express u^{*i}_{kl} as shown:

$$u^{*i}_{kl} = G_{kl,jj} - \frac{1}{2(1-\nu)} \, G_{kj,lj} \qquad (5.96)$$

The Galerkin tensor [92, 82, 34] for two-dimensional problems is given by

$$G_{kl} = \frac{1}{8\pi\mu} \, (r^i)^2 \, \ln\left(\frac{1}{r^i}\right) \, \delta_{kl} \qquad (5.97)$$

A remark must be made here. The fundamental solution which is generated by substituting the Galerkin tensor (5.97) into equation (5.96) is

$$\begin{aligned}
u^*_{kl} &= \frac{1}{8\pi(1-\nu)\mu} \Big[(3-4\nu)\ln\left(\frac{1}{r}\right)\delta_{kl} - \\
&\quad \left(\frac{7-8\nu}{2}\right)\delta_{kl} + r_{,k}\,r_{,l} \Big]
\end{aligned} \qquad (5.98)$$

This equation differs from the previous fundamental solution by a constant. This difference may be interpreted as a rigid body translation , and as such, should not affect the final results. Nevertheless, to be consistent, the fundamental solution given in equation (5.98) will be used when applying the Galerkin tensor to take the domain integrals to the boundary.

Substituting (5.96) into the domain integral in equation (5.95), after further manipulations, yields:

$$\int_\Omega \psi \, u^{*i}_{kl,l} \, d\Omega = \frac{1-2\nu}{2(1-\nu)} \int_\Omega \psi \, \nabla^2(G_{kl,l}) \, d\Omega \qquad (5.99)$$

Green's second identity [66] can be conveniently applied, together with expression (5.92), to rewrite equation (5.99) as

$$\int_{\Omega} \psi \, u^{*i}_{kl,l} \, d\Omega = \frac{1-2\nu}{2(1-\nu)} \left\{ \int_{\Gamma} G_{kl,l} \, \psi_{,j} \, n_j \, d\Gamma - \right.$$

$$\left. \int_{\Gamma} G_{kl,lj} \, \psi \, n_j \, d\Gamma - K_0 \int_{\Gamma} G_{kl} \, n_l \, d\Gamma \right\} \qquad (5.100)$$

Now taking equation (5.100) into (5.95) gives the component B_i in terms of boundary integrals only, as follows:

$$\int_{\Omega} \psi \, u^{*i}_{kl,l} \, d\Omega = \int_{\Gamma} \psi \, u^{*i}_{kl} \, n_l \, d\Gamma + \frac{1-2\nu}{2(1-\nu)} \left\{ \int_{\Gamma} G_{kl,l} \, \psi_{,j} \, n_j \, d\Gamma - \right.$$

$$\left. \int_{\Gamma} G_{kl,lj} \, \psi \, n_j \, d\Gamma - K_0 \int_{\Gamma} G_{kl} \, n_l \, d\Gamma \right\} \qquad (5.101)$$

According to the type of body force under consideration the appropriate expressions for ψ and K_0 should be substituted in the above equation. As an example, the case of gravity loads will be shown here.

Gravitational loads

In this case the potential ψ is given by

$$\psi = -\rho \, g \, x_2 \qquad (5.102)$$

hence its Laplacian is a constant, i. e.

$$\nabla^2 \psi = 0 \qquad (5.103)$$

where ρ is the mass density and g is the gravitational acceleration, here supposed to act in the x_2 direction. Both are assumed to be constant.

After substituting the proper expressions for ψ and K_0 into equation (5.101), the domain integral associated to a constant gravitational field can be written as

$$B_i = \int_{\Omega} u^{*i}_{kl} \, b_l \, d\Omega = \int_{\Gamma} u^{*i}_{kl} \, \rho \, g \, x_2 \, n_k \, d\Gamma +$$

$$\frac{(1 - 2\nu)\,\rho\,g}{16\pi\,\nu(1 - \nu)} \int_{\Gamma} \left(r_{,k}\,n_2 - x_2\,r_{,k}\,n_j \right)\,d\Gamma \qquad (5.104)$$

The cases of a thermal loading due to a steady-state temperature field and the body force resulting from a constant rigid body rotation about a fixed axis could be similarly treated. These cases are presented in detail in references [79, 64, 65].

Chapter 6

Numerical Applications

6.1   Introduction

The hybrid-displacement boundary element formulation, which has been proposed in the earlier parts of this work, is a novel numerical technique. Several problems in two-dimensional potential and elasticity theories have been solved to demonstrate its feasibility and accuracy. Some of these solutions will be presented in this chapter and their accuracy will then be assessed by comparing these numerical results to analytical solutions. Hybrid boundary elements results are also compared to numerical results obtained by the conventional boundary element method. The performance of the new approach can then be compared against the performance of the conventional BEM, which has already been proved to be efficient and accurate. The convergence of the hybrid-displacement method has also been demonstrated numerically and the results are shown for some of the cases studied in function of an error norm.

In order to study the convergence of the solution the following error norm is defined:

$$\epsilon = \frac{1}{P}\sqrt{\frac{\sum_{i=1}^{P}(v_i - \overline{v}_i)^2}{\sum_{i=1}^{P}(\overline{v}_i)^2}} \tag{6.1}$$

where v_i can represent either the numerical solution for the potential u or for its normal derivative $\partial u/\partial n$ at the point i; $\overline{v}_i$ can be the analytical solution for either the potential or its normal derivative at the point i and P is the total number of points considered.

This error norm will only be used for studies of convergence. The other measure of the error, which is denoted simply by *Error*, is defined as the difference between the numerical and exact solutions. It can be given as percentage of the exact solution; in this case it is called *Error %*.

In the cases tested it has been demonstrated that the equilibrium for fluxes on the boundary (potential problems) or for tractions (elasticity) is satisfied in a variational sense.

6.2 Examples for Potential Problems

In this section, both constant and isoparametric quadratic hybrid boundary elements are applied to the solution of two-dimensional potential problems.

6.2.1 Constant Elements

Example 1

As a first numerical test the simple case of a uni-dimensional heat flow in a square domain subject to the mixed boundary conditions shown in figure 6.1 is considered. The generated potential field is linear.

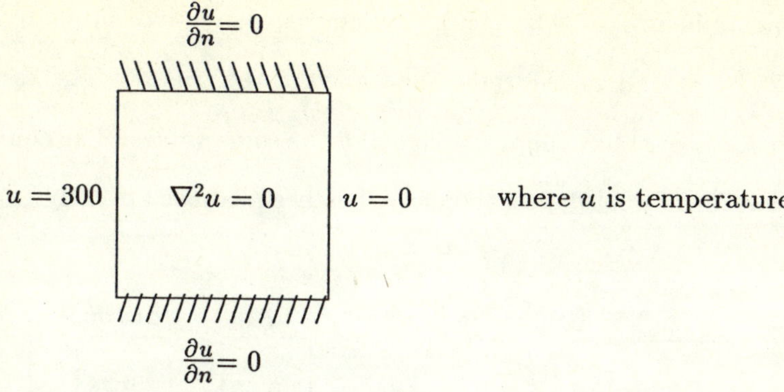

Figure 6.1: Heat flow in a square domain: boundary conditions

This problem has also been solved by the conventional direct boundary element method [40]. Numerical results for both methods, the hybrid and the conventional BEM, are presented in figure 6.2. They correspond to a mesh of 16 constant boundary elements of equal length.

The discontinuities in the flux at the corners can be represented by constant elements. This occurs because the constant element is a discontinuous element, i. e. the approximate potential and flux on the boundary present discontinuities as they can vary stepwise from element to element.

In this example the results obtained using the hybrid boundary element, particularly for potentials and fluxes on the boundary, are more accurate than the ones obtained with the conventional boundary elements.

The convergence of the solution has been studied by using the error previously defined. Four different meshes containing 8, 16, 32 and 64 constant elements have been used. The results in fig 6.3 demonstrates that the convergence is excellent and the errors rapidly tend to zero as the number of elements increases. The results for normal derivatives presented in figure 6.2 correspond to the right-hand edge of the square. The results for the left-hand edge have opposite sign and the same values. The equilibrium condition for normal fluxes, therefore, has been satisfied exactly.

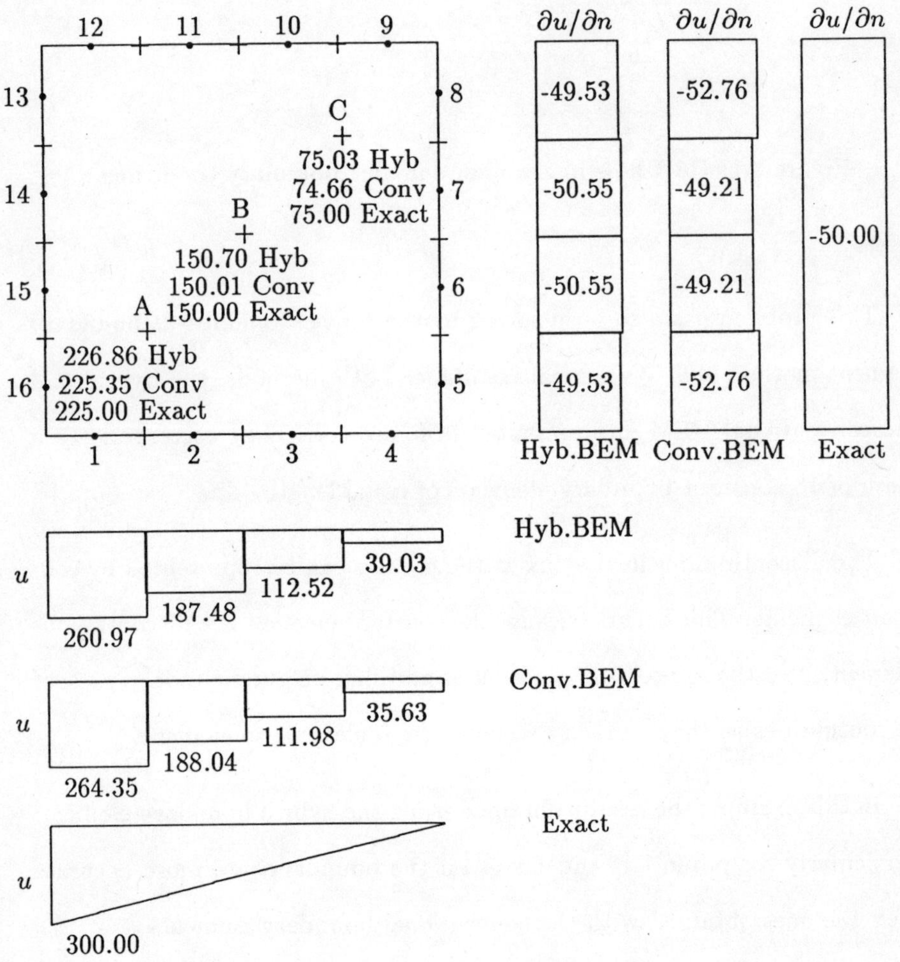

Figure 6.2: Solutions on the boundary and at internal points in example 1

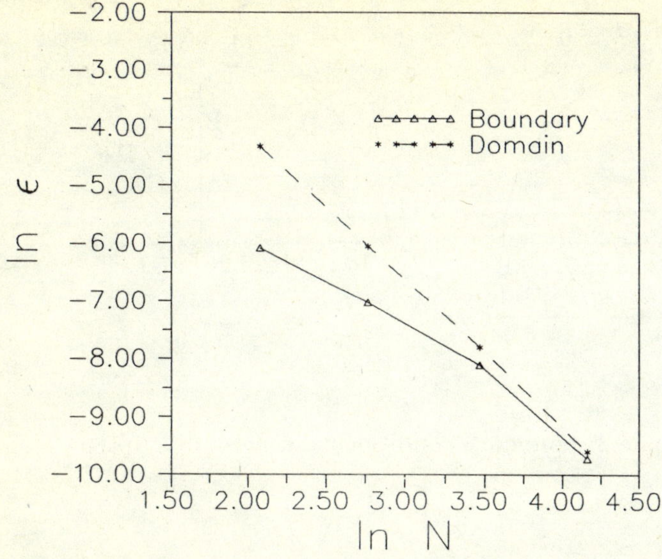

Figure 6.3: Convergence of results in example 1

Example 2

The example studied is the problem of heat flow in a semicircular domain, as shown in figure 6.4. This case is interesting because the exact solution for fluxes presents a singularity at point O, where the potential u is discontinuous. The region of interest is a semicircular domain with zero flux at a distance $r = 30$ from the origin. The analytical solution for this problem, which is shown in figure 6.5, is presented in reference [85].

Results for temperatures on the boundary and at four internal points are given in table 6.1. In this particular case, the hybrid solution is more accurate than the conventional boundary elements solution for points on the boundary. For internal points, however, the conventional BEM solution is shown to be more precise.

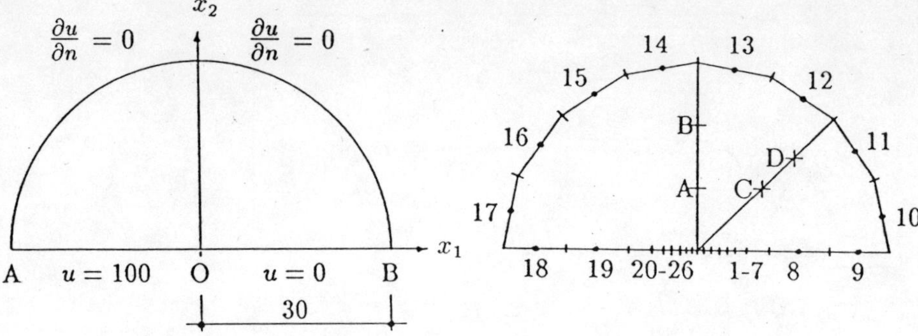

Figure 6.4: Example 2: boundary conditions and mesh description

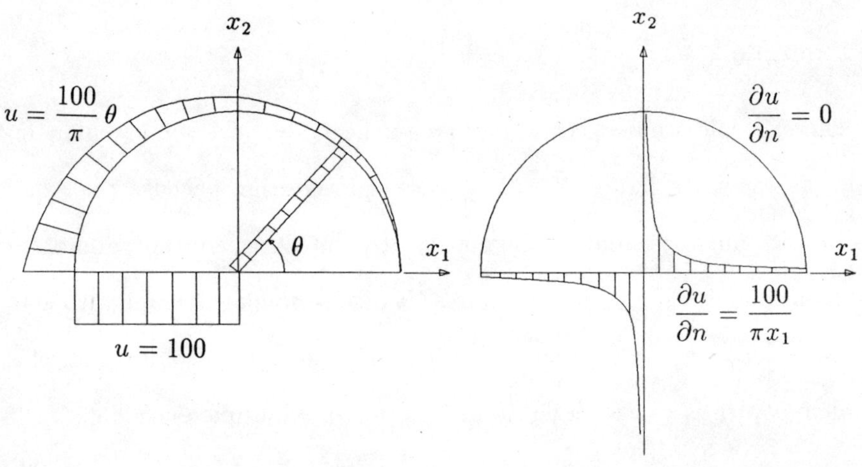

Figure 6.5: Example 2: analytical solution

Point	Hybrid BEM		Conventional BEM		Exact
	u	Error %	u	Error %	u
10	6.49213	+3.87	5.84429	-6.49	6.25000
11	18.68818	-0.33	18.52718	-1.19	18.75000
12	31.21732	-0.10	31.14179	-0.35	31.25000
13	43.73918	-0.02	43.72010	-0.07	43.75000
A	50.10009	+0.20	50.00308	+0.01	50.00000
B	50.14477	+0.29	50.00545	+0.01	50.00000
C	25.06197	+0.25	24.97533	-0.10	25.00000
D	25.01610	+0.06	24.90107	-0.40	25.00000

Table 6.1: Results for temperatures on the boundary and at internal points

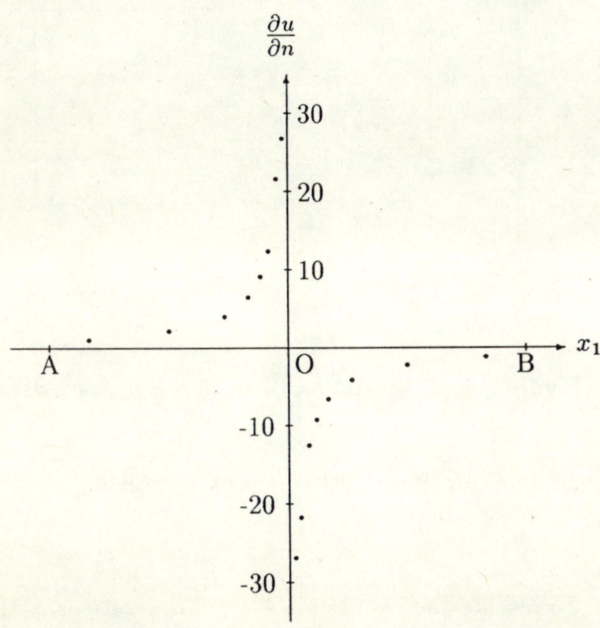

Figure 6.6: Distribution of fluxes along the x_1 axis.

The variation of normal flux along $x_2 = 0$, as plotted in figure 6.6, clearly shows the presence of the singularity in fluxes.

Example 3

A Dirichlet problem in a L-shaped domain (figure 6.7) is studied as a further application. This problem consists of determining the potential u in the domain and its normal derivative on the boundary, given that $u = x$ on the boundary. The analytical solution is $u = x$ throughout the domain. This example was presented by Jaswon and Symm in reference [61].

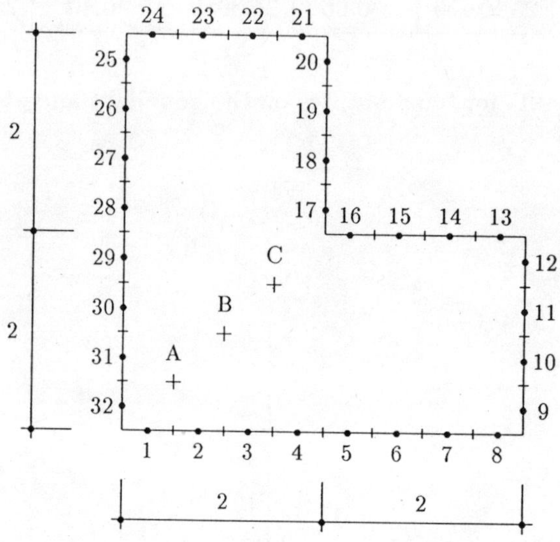

Figure 6.7: Potential problem in a L-shaped domain

A mesh of 32 constant elements have been used to assess the accuracy of the method.

Table 6.2 presents the results for normal derivatives of the potential on the boundary, which are very accurate when compared against the exact solution and are generally better than those obtained using the conventional

	$\partial u / \partial n$		
Point	Hybrid BEM	Conventional BEM	Exact Solution
1	-0.03058	0.04784	0.00000
3	-0.00093	0.00004	0.00000
4	-0.00007	-0.00232	0.00000
11	1.00166	0.98455	1.00000
12	0.98950	1.08680	1.00000
13	0.01175	-0.4918	0.00000
16	0.00076	-0.04720	0.00000
17	1.00179	1.08909	1.00000
24	-0.03644	0.04632	0.00000
25	-0.96637	-1.08856	-1.00000
32	-0.97208	-1.08815	-1.00000

Table 6.2: Normal derivatives of the potential on the boundary

BEM. It is interesting to notice that the re-entrant corner is not the region where the greatest error occurs. This fact has been pointed out and explained by Jaswon and Symm in reference [61]. For this particular discretization, the source of the greatest error is the top left corner (elements 24 and 25). At the re-entrant corner (elements 16 and 17) the solution is very accurate and much better than the conventional direct BEM solution.

Table 6.3 shows the values of potential at internal points A, B and C (figure 6.7). These results are also accurate although the error for point A is larger than the one found when using the classical direct BEM. This increase in the error has been attributed to being the point A near to the boundary. Indeed, the proximity of a boundary node can cause a considerable error in the solution for potentials at internal points. This can be understood by taking into consideration formula (2.51), which is used to evaluate the internal potentials: it is singular at a boundary node. For points B and C the hybrid BEM solution is much more accurate than the solution found by the conventional direct BEM.

Point	Potential u				
	Hybrid BEM	Error %	Conventional BEM	Error %	Exact
A	0.49592	-0.82	0.50038	+0.08	0.50000
B	0.99975	-0.03	1.00194	+0.19	1.00000
C	1.50135	+0.09	1.50535	+0.36	1.50000

Table 6.3: Results for potentials at internal points

The convergence of the results have also been studied by using 8, 16, 32 and 64 elements of equal length. Figure 6.8 shows that the convergence for

both solutions, on the boundary and at internal points, present a good rate
of convergence. This convergence is more rapid in the case of the solution
inside the domain.

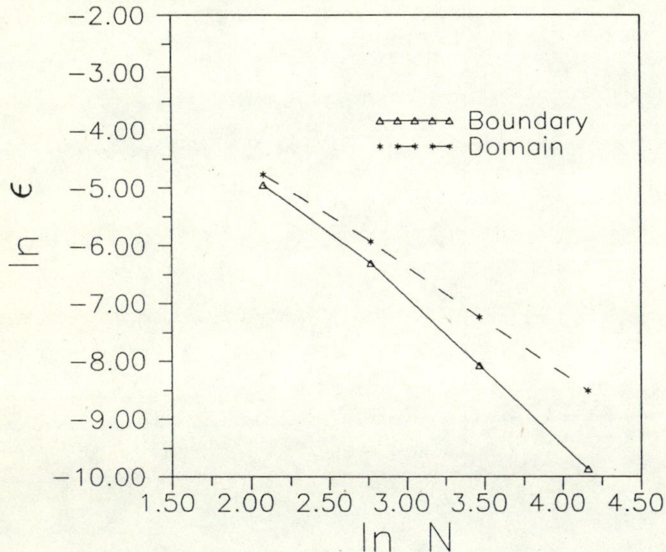

Figure 6.8: Convergence of the solution in example 3

6.2.2   Quadratic Elements

Example 4

As a first example with quadratic elements, a Dirichlet problem defined
in the circle of radius 1 and centred at the origin $(x^2 + y^2 \leq 1)$ is considered.
The potentials on the boundary are given by $u = -x^3 - 3x^2y + 3xy^2 + y^3$.
The analytical solution of the problem is $u = -x^3 - 3x^2y + 3xy^2 + y^3$ in Ω.

The example is a good test for quadratic elements because the boundary is curved, the potential field presents a cubic variation and the normal derivatives vary quadraticaly.

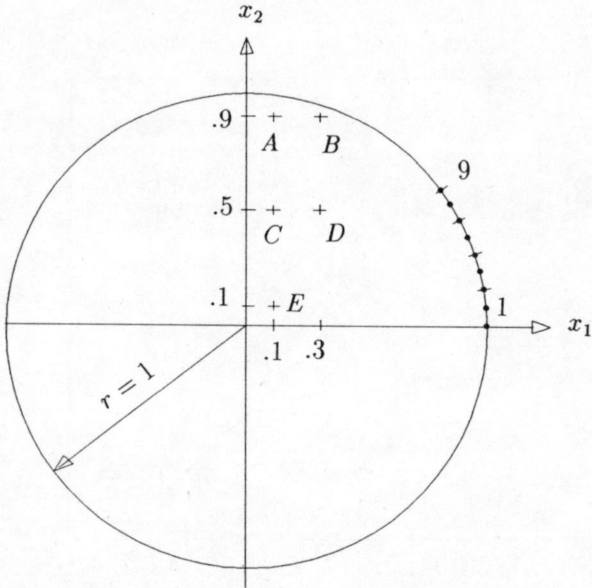

Figure 6.9: Example 4: boundary nodes and internal points

This problem was taken from reference [86] where it was used as a test problem for the classical direct boundary element method. It has been also previously solved in [87, 88], using different numerical techniques.

The results presented in tables 6.4 and 6.5 correspond to a mesh of 32 boundary elements and to the internal points depicted in figure 6.9. The solution on the boundary (normal derivatives) is shown to be accurate when compared to the analytical solution and it is much better than the solution obtained by the conventional direct BEM.

It has already been mentioned in this chapter that the proximity of a boundary node can cause considerable errors in the solution at internal points. This fact can be clearly seen in table 6.5 where the solution for points A and B are not as accurate as the solution for the other points.

Point	Normal Derivatives of the Potential, $\partial u/\partial n$				
	Exact	Hybrid BEM	Error	Conventional BEM	Error
1	-3.0000	-3.0322	-0.0322	-3.3622	-0.3622
3	-4.1611	-4.2056	-0.0445	-4.5239	-0.3628
5	-3.9197	-3.9628	-0.0431	-4.2841	-0.3644
7	-2.3571	-2.3835	-0.0264	-2.7209	-0.3638
9	0.0000	0.0000	0.0000	-0.3638	-0.3638

Table 6.4: Results for $\partial u/\partial n$ on the boundary

Point	Potentials at internal points				
	Exact	Hybrid BEM	Error	Conventional BEM	Error
A	0.9440	0.9463	+0.0023	0.9440	0.0000
B	1.1880	1.1940	+0.0060	1.1880	0.0000
C	0.1840	0.1840	0.0000	0.1840	0.0000
D	0.1880	0.1880	0.0000	0.1880	0.0000
E	0.0000	0.0000	0.0000	0.0000	0.0000

Table 6.5: Results for potential at internal points

The numerical convergence of the results for this example have also been studied. Four meshes of 4, 8, 16 and 32 quadratic elements of equal length have been used and the results are shown in figure 6.10. It can be seen that the results present a good convergence, similarly to the constant element case.

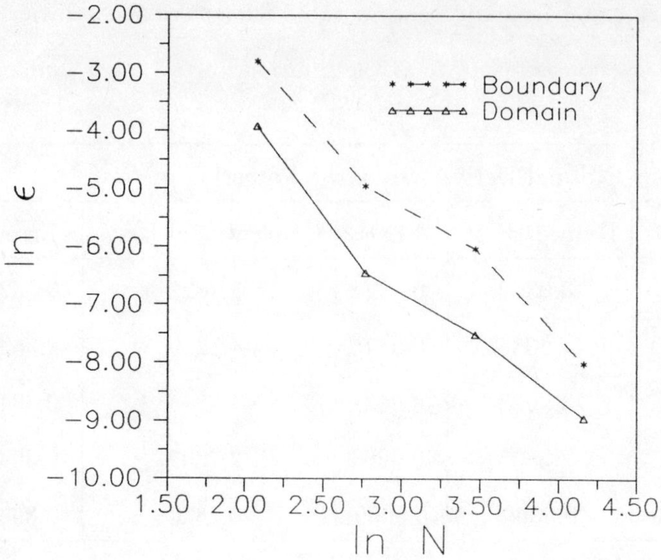

Figure 6.10: Convergence of the solution in example 4

Example 5

Another distribution of temperatures in a square domain of side $a = 6$ will be studied. In this case, the prescribed boundary variables are given in figure 6.11. The corresponding analytical solution is presented in [85] and is shown in figure 6.12.

The results for boundary fluxes along the x_1 axis are shown in table 6.6. Five internal points (figure 6.11) equally spaced over the diagonal of the square have been chosen to assess the accuracy of the solution inside the domain. The results for these points are shown in table 6.7. In these tables, the results correspond to a mesh of 16 boundary elements of equal length.

The numerical convergence of the results have been studied by using four meshes with 4, 8, 16 and 32 elements of equal length. The results are shown in figure 6.13.

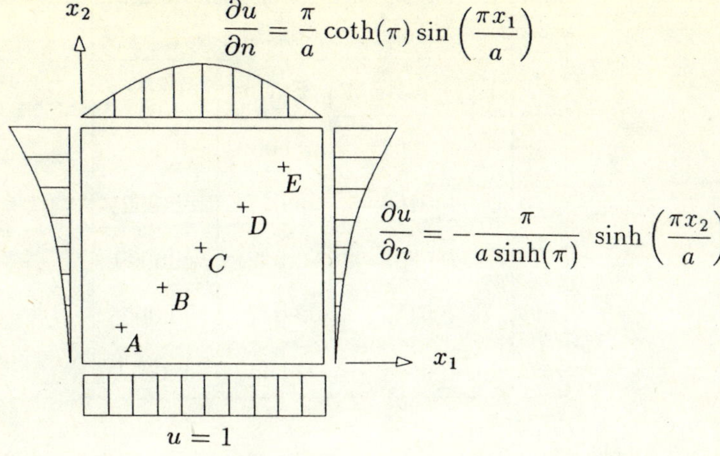

Figure 6.11: Example 5: boundary conditions and internal points

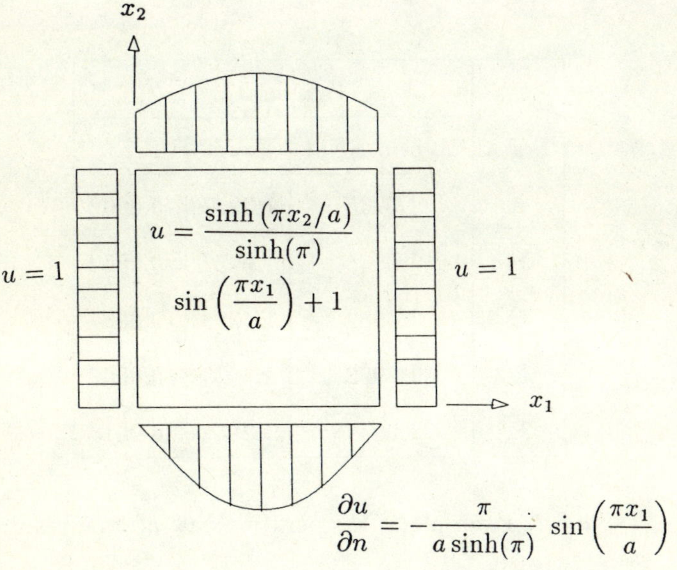

Figure 6.12: Example 5: analytical solution

x_1	Flux		
	Hybrid BEM	Exact	Error
1	-0.0002	0.0000	-0.0002
2	-0.0169	-0.0174	+0.0005
3	-0.0332	-0.0321	-0.0009
4	-0.0413	-0.0419	+0.0006
5	-0.0464	-0.0453	-0.0011

Table 6.6: Example 5: variation of fluxes along the x_1 axis

Point	Temperature		
	Hybrid BEM	Exact	Error
A	1.0235	1.0237	-0.0002
B	1.0935	1.0937	-0.0002
C	1.1989	1.1993	-0.0004
D	1.2992	1.2999	-0.0007
E	1.2939	1.2952	-0.0013

Table 6.7: Example 5: temperatures at internal points

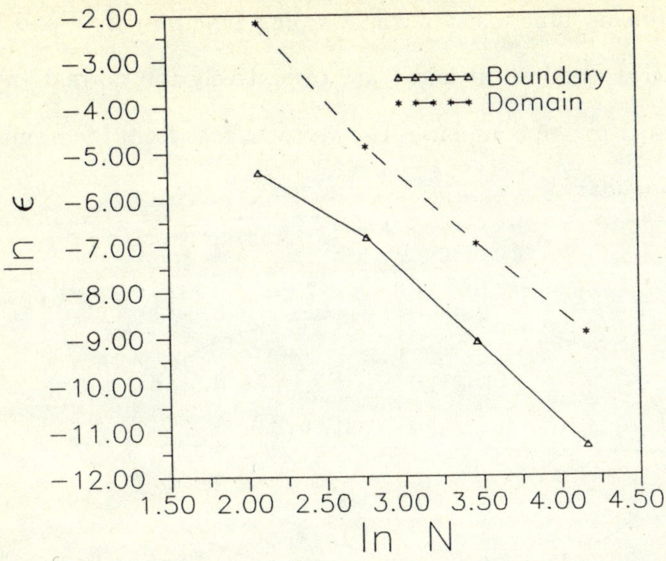

Figure 6.13: Convergence of the solution in example 5

Example 6

As the last example in potential theory, the problem of a prismatic
bar under torsion is solved by the hybrid-displacement boundary element
method. The bar has an elliptical cross-section with half minor and major
axis a and b, respectively. By supposing that the hypotheses of St. Venant's
theory of torsion apply, the solution of the problem is given by the following
boundary-value problem:

$$\nabla^2 \phi = 0 \qquad \text{in } \Omega \tag{6.2}$$

with the boundary condition

$$\frac{\partial \phi}{\partial n} = r \cos(\mathbf{r}, \mathbf{t}) \qquad \text{on } \Gamma \tag{6.3}$$

where the harmonic function ϕ is the warping function; **n**, **r** and **t** are defined for a cross-section. **n** and **t** are respectively the normal and the tangential vectors to the boundary. The vector **r** goes from the origin to a point on the boundary.

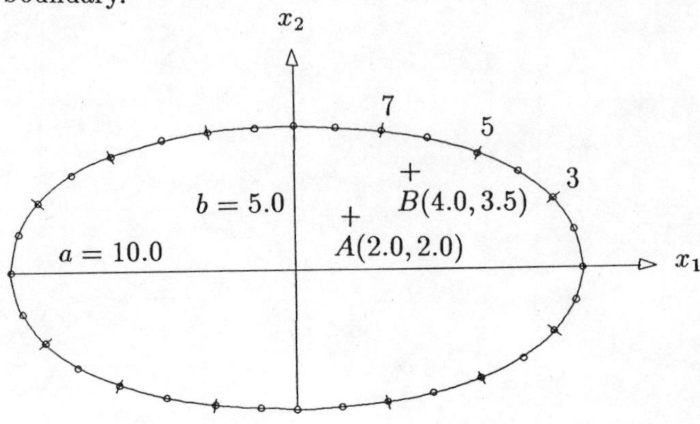

Figure 6.14: Elliptical section under torsion

The displacements can then be found in terms of ϕ and the torsion angle per unit length θ, as follows:

$$u_1 = -\theta\, x_2\, x_3 \tag{6.4}$$

$$u_2 = \theta\, x_1\, x_3 \tag{6.5}$$

$$u_3 = \theta\, \phi \tag{6.6}$$

where x_1 and x_2 are parallel to the principal axis and x_3 is parallel to the axis of the bar (figure 6.14); $\theta = T/GJ$; T is the applied torsional moment; G is the shear modulus; J is an "effective" polar moment of area, which for an elliptical section is given by

$$J = \frac{\pi a^3 b^3}{a^2 + b^2} \tag{6.7}$$

This Neumann problem has a solution since the condition

$$\int_\Gamma \frac{\partial \phi}{\partial n} \; d\Gamma = 0 \tag{6.8}$$

is fulfilled by the prescribed boundary conditions. For this solution to be unique the value of ϕ at one point, at least, should be given. This can be done by using the symmetry that exists for the problem.

This example has been taken from reference [65] where it was solved by the conventional BEM, also using quadratic boundary elements. Herein the boundary has been subdivide into 16 quadratic elements of equal length (figure 6.14). The results presented in table 6.8 are accurate in comparison to the exact solution. They also show good agreement to the direct BEM solution, although the conventional BEM solution is more accurate.

Point	Hybrid BEM	Error	Conventional BEM	Error	Exact
		Warping function ϕ			
3	-12.443	+0.041	-12.506	-0.022	-12.484
5	-14.548	+0.022	-14.576	-0.006	-14.570
7	-9.376	-0.020	-9.363	-0.007	-9.356
A	-2.396	+0.004	-2.399	+0.001	-2.400
B	-8.418	-0.018	-8.403	-0.003	-8.400

Table 6.8: Results for ϕ on the boundary and at internal points

6.3 Elasticity Problems

In this section, the new formulation proposed in the previous part of this work is tested for the case of elastostatics problems. The adequacy of the method to a computer implementation will be demonstrated. The two types of elements which have been proposed in chapter 5 (constant and isoparametric quadratic elements) will be tested, by assessing the accuracy and the convergence of the solutions obtained when using these elements.

6.3.1 Constant Elements

Three classical elasticity problems are solved by the present approach with constant elements. The results are compared to both conventional BEM and analytical solutions.

Example 7

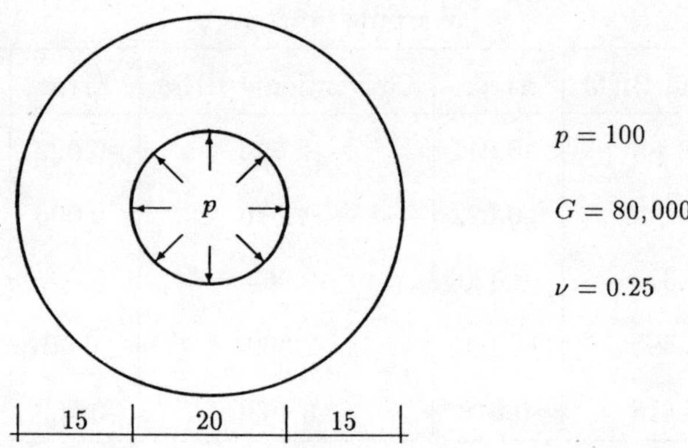

Figure 6.15: Hollow cylinder under internal pressure

The case of a hollow cylinder under internal pressure is presented as a first numerical test in elastostatics (figure 6.15). This problem has been used by several authors [40, 65] as a test problem and an analytical solution is available [89]. Due to the symmetry of the problem, only one quarter of the cylinder need to be modelled.

In order to assess the accuracy and the convergence of this hybrid formulation, in the case of constant elements in elasticity problems, three different meshes have been used. The first contains 15 boundary elements as shown in figure 6.16. The other meshes are obtained by successively doubling the number of elements. Elements of same length have been used in each part of the boundary.

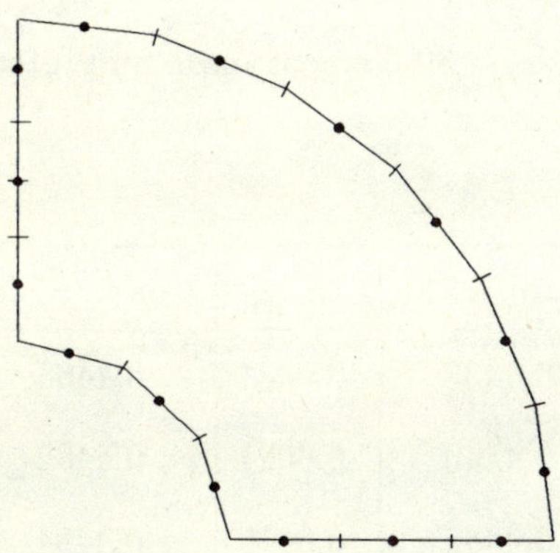

Figure 6.16: Hollow cylinder under internal pressure: 15 element mesh

The hybrid BEM solution is compared to the conventional BEM solution for the mesh with 30 elements shown in figure 6.17. The radial distances for points A, B and C are respectively 13.75, 17.5 and 21.25. The results for the radial displacements at internal points are presented in table 6.9 and for circumferential stresses in table 6.10.

The results for displacements and stresses are accurate when compared to the analytical solution. In comparison to conventional BEM results, the

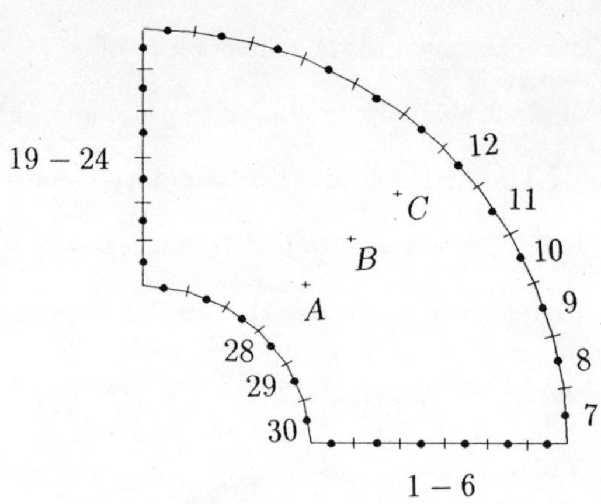

Figure 6.17: Example 7: 30 element mesh and internal points

Point	Exact	Hyb. BEM	Error	Conv. BEM	Error
7	0.4464	0.4433	-0.0031	0.4468	0.0004
8	0.4464	0.4440	-0.0024	0.4482	0.0018
9	0.4464	0.4445	-0.0019	0.4494	0.0030
10	0.4464	0.4448	-0.0016	0.4504	0.0040
11	0.4464	0.4450	-0.0014	0.4510	0.0046
12	0.4464	0.4451	-0.0013	0.4513	0.0049
28	0.8036	0.8034	-0.0002	0.8266	0.0230
29	0.8036	0.8021	-0.0015	0.8268	0.0232
30	0.8036	0.8030	-0.0006	0.8251	0.0215
A	0.6230	0.6208	0.0022	0.6319	0.0089
B	0.5294	0.5275	0.0019	0.5374	0.0080
C	0.4766	0.4752	0.0015	0.4838	0.0072

Table 6.9: Radial displacements for hollow cylinder ($\times 10^2$)

Point	Exact	Hyb. BEM	Error	Conv. BEM	Error
A	82.01130	81.7240	-0.2875	82.0192	+0.0079
B	57.9226	57.8500	-0.0727	58.1691	+0.2465
C	45.4112	45.4849	+0.0738	45.6575	+0.2463

Table 6.10: Circumferential stresses at internal points

hybrid solution is, in general, more accurate for the displacements both on the boundary and at internal points. For the boundary nodes depicted in figure 6.17, the hybrid BEM results are more accurate on 10 among 12 points, showing considerable superiority in relation to the conventional BEM. In the case of stresses, the hybrid BEM results for the internal points used are also more accurate: it gives better results for two of the three internal points, as shown in table 6.10.

Example 8

The second example considered here is the classical stress concentration problem of a square plate of width $2l$, with a small central circular hole of radius a (figure 6.18). The plate is subjected to a uniform tensile stress σ in the x_2 direction. The ratio of the diameter of the hole to the width of the plate is taken to be $a/l = 0.1$ and the Poisson's ratio for the material is $\nu = 0.25$.

Figure 6.19 shows the variation of normal stresses in x_2 direction along the x_1 axis with distance from the hole. Both the conventional BEM and the hybrid BEM solutions were obtained by using the same 29 boundary constant elements mesh, shown in figure 6.18. K_t is the stress concentration factor defined as $K_t = \sigma_{22}/\sigma$.

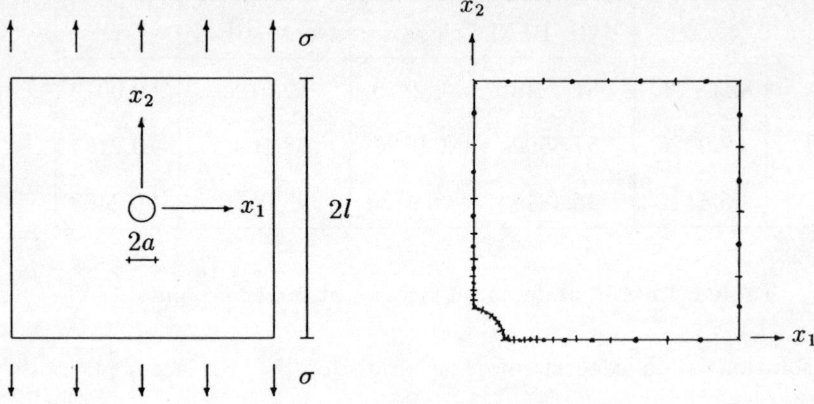

Figure 6.18: Example 8: definition of the problem and discretization

The analytical solution curve corresponds to the case of an infinite plate and is taken from Timoshenko [89] which gives a stress concentration factor of 3.000. This value increases slightly to 3.024 for the finite width plate [90].

The results for the hybrid and conventional BEM are in good agreement and they are both accurate when compared with the exact solution. Although the constant elements used here are not appropriate for modelling curved boundaries and problems where the solution varies rapidly, the numerical results follows the analytical curve and simple extrapolation of the boundary element values give an accurate estimate of the stress concentration factor.

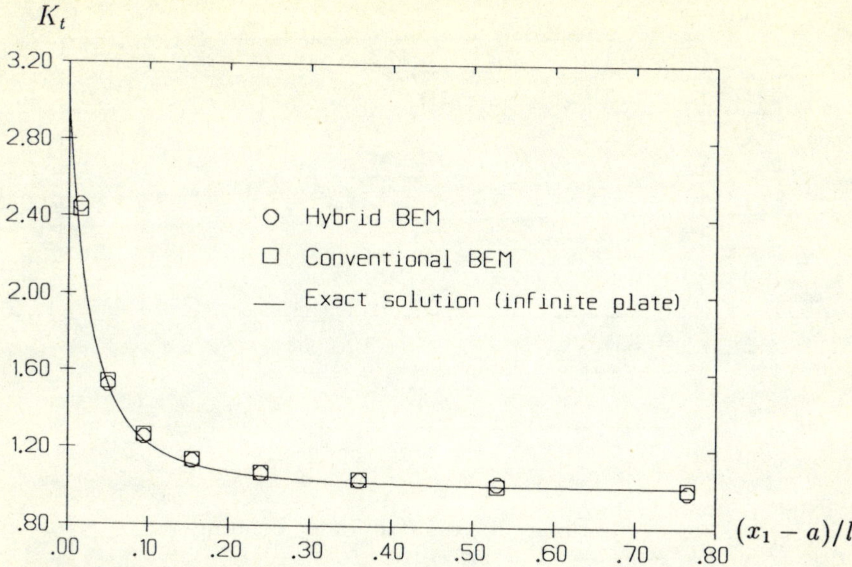

Figure 6.19: Variation of stresses with distance from the hole

Example 9

The following example will show that constant elements, in this hybrid approach, can give reasonable results even for problems involving the bending of plates. However, to reach a good accuracy a great number of elements should be used.

The example consists of a rectangular plate under a linear distribution of tractions such that their resultant is equivalent to two opposite applied moments (figure 6.20). The elastic properties of the material are: shear modulus $G = 80000$ and Poisson's ratio $\nu = 0.25$.

The plate is considered to be in plane stress and therefore the equivalent Poisson's ration ν' as given in formula (5.1) should be used.

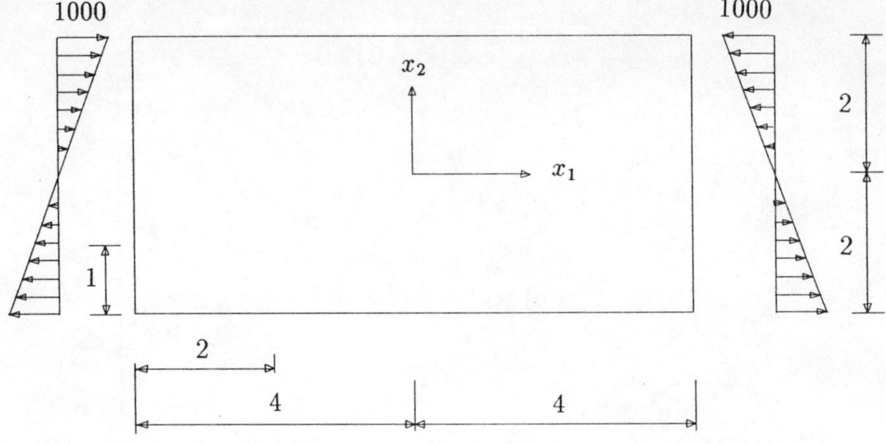

Figure 6.20: Plate under bending

This problem was used in reference [65] to show the accuracy of the quadratic elements in the conventional BEM. It is also shown that poor results are obtained using constant elements in spite of the use of a great number of elements: errors of 21.5% and 17.5% for, respectively, u_1 and u_2 are reported corresponding to a mesh of 50 constant elements (the whole plate).

In table 6.11 the results for displacement of the corner A (using constant elements in the hybrid-displacement BEM) are shown to converge towards the exact solution . These results are also shown to be much more accurate than the conventional BEM results for the same discretization. The results for u_1 at point A were obtained by using linear extrapolation of the results for u_1 on the two top elements, which are close to the corner A. Only one half of the plate was discretized and the symmetry conditions were imposed as displacement boundary conditions. Elements of equal size were used for all the meshes.

Displ. component	Exact solution	16 element mesh		32 element mesh	
		Hyb.BEM	Conv.BEM	Hyb.BEM	Conv.BEM
u_1	-0.020	-0.0169	-0.0135	-0.0195	-0.0168
u_2	0.020	0.0186	0.0136	0.0204	0.0170

Table 6.11: Plate under bending: displacement at corner A

6.3.2 Quadratic Elements

Example 10

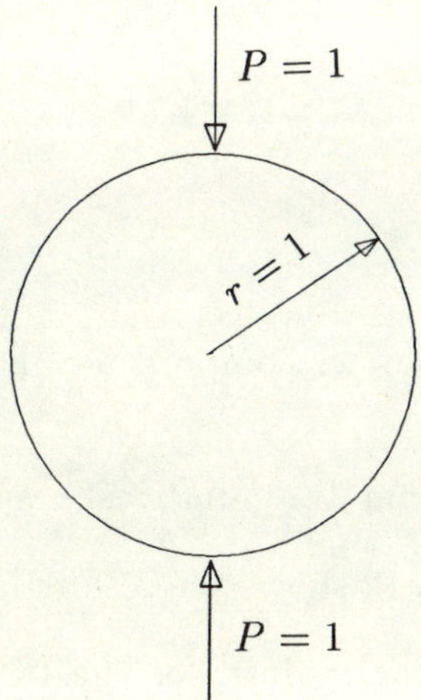

Figure 6.21: Cylinder under compression: definition of the problem

The problem of a cylinder under compression will now be presented. The length of the cylinder is suppose to be big enough for a cross section (at a reasonable distance from the extremities) to be considered in plane strain.

The exact solution for the stress field is presented in reference [89]. The convergence for stresses at the internal points shown in figure 6.22 have been studied. Four meshes have been used with 4, 8, 16 and 32 equal elements, respectively.

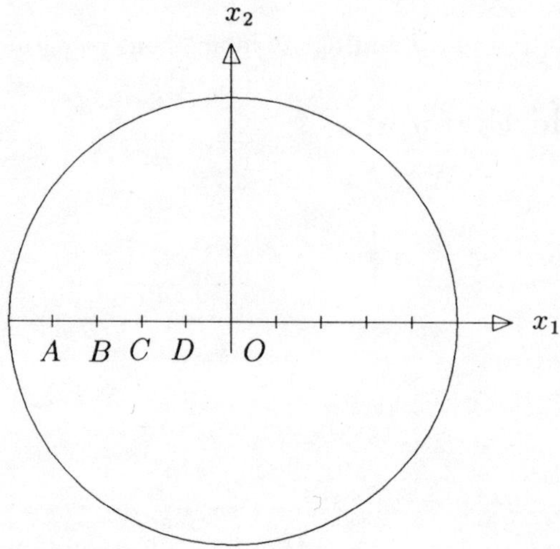

Figure 6.22: Cylinder under compression: internal points

Figure 6.23 demonstrates that the numerical results converge towards the analytical solution. Table 6.12 shows the results for the normal stresses in the x_2 direction at the internal points, for the 16 element mesh, which present a good degree of accuracy. Once more it is possible to see that the results for the internal points, when they are close to the singularities (boundary nodes), are less accurate than the results for points away from the singularities.

In this example, the concentrated loads have been introduced as boundary conditions. This is an interesting feature of this hybrid BEM formulation: it allows concentrated loads to be considered as such. In the conventional BEM formulation only distributed loads can be used as boundary conditions.

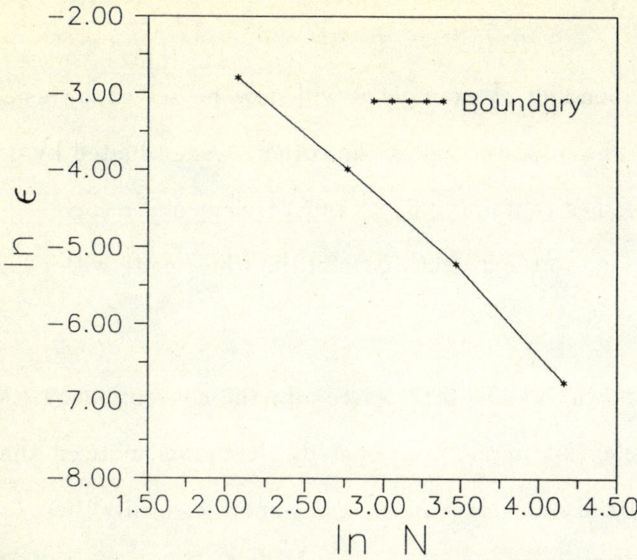

Figure 6.23: Example 10: Convergence of the stresses σ_{22} at internal points

Point	Exact	Hyb. BEM	Error %
A	-0.15551	-0.1579	+1.8
B	-0.3701	-0.3733	+0.9
C	-0.6279	-0.6281	+0.0
D	-0.8588	-0.8549	+0.5
O	-0.9549	-0.9490	+0.6

Table 6.12: Example 10: stresses σ_{22} at internal points

Example 11

The plate under bending of example 9 will now be solved by using quadratic elements. The displacement at the corner A is evaluated by using three different meshes containing 6, 12 and 24 elements, respectively. These meshes correspond to the discretization of the whole plate with equal elements.

The results are shown in table 6.13 where also the conventional BEM solution (for the 6 element mesh) is presented. It can be noticed that an accurate hybrid BEM analysis requires many more elements than the conventional direct BEM.

Displacement component	Analytical solution	Conv. BEM 6 elements	Hybrid. BEM		
			6 elements	12 elements	24 elements
u_1	-0.0200	-0.0200	-0.0196	-0.0199	-0.0200
u_2	0.0200	0.0200	0.0186	0.0197	0.0200

Table 6.13: Displacement at corner A

It is interesting to point out here that the hybrid solution for this example using constant elements is much better than the conventional BEM solution whereas in the case of quadratic elements the opposite occurs, as commented in the previous paragraph.

Example 12

As the last example, the solution of the cantilever beam, which is shown in figure 6.24, will be presented. It can be considered a plane stress problem in which the elastic constants are assumed to be the same as those in example 9, i.e. $G = 80000$ and $\nu = 0.25$.

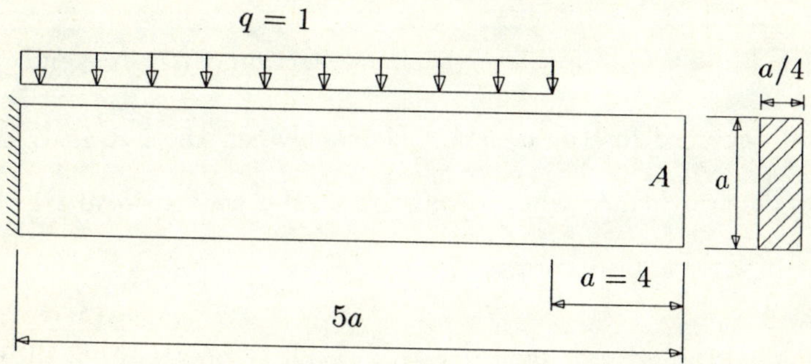

Figure 6.24: Cantilever beam

The hybrid BEM results for the vertical displacement at the free end (point A) is presented in table 6.14, where it is compared against the conventional BEM results. The elastic beam theory gives for this problem a free end deflection of

$$u_2^A = -\frac{128}{3EI}\, qa^4 = 0.0102 \tag{6.9}$$

Since the ratio between the depth of the beam and its length is equal to 5, there is a small contribution in the total deflection due the shear which is not considered in the beam theory.

When solving bending problems with quadratic elements, the hybrid-displacement BEM requires a more refined mesh than the conventional BEM, to give the same accuracy. The results show that the quadratic element is a bit stiff. This behaviour was unexpected because an opposite

Displacement	12 elements		24 elements	
component	Hyb. BEM	Conv. BEM	Hyb. BEM	Conv. BEM
u_2^A	0.0881	0.0105	0.0103	0.0106

Table 6.14: Cantilever beam: deflection at the free end

situation occurred for the constant element when the hybrid BEM had a very good performance when compared to the conventional BEM, in the bending of a plate (see example 9).

In the hybrid finite element theory [1, 91] it is reported that the elements can be made more flexible if polynomials of lower order than the ones used for the displacement were used to represent the tractions. Based on this statement, in the case of this hybrid BEM formulation, one believes that the results for the quadratic element, could be improved by using a lower order polynomial to expand the tractions. Another possibility for improving the results is to use discontinuous elements, as it is done in the conventional BEM. In that method, it is known that for bending problems discontinuous elements are more accurate than the continuous ones. Both alternatives, however, are out of the scope of this work.

Chapter 7

Conclusions

In this work, a novel boundary element formulation has been presented. It has been derived and implemented computationally to solve problems in potential theory and linear elastostatics.

The new approach has been called the hybrid-displacement boundary element method due to its two main characteristics: it is based on a multi-field variational principle and it is a boundary formulation.

The functional that is used in the generation of the formulation is derived from classical one-field functionals in both the potential and the elasticity cases. Different independent field variables are assumed on the boundary and inside the domain (potentials in potential problems and displacements in elastostatics). The compatibility condition between them is then introduced in the functional by means of Lagrange multipliers. These are later identified as another boundary variable (fluxes in potential and tractions in elasticity problems). The resulting functional, therefore, involves three independent field variables: two of them defined on the boundary and the other in the domain. The reason why the proposed formulation is called a hybrid-displacement formulation is because the generation of the hybrid-displacement finite element models follows the same pattern.

The formulation becomes a boundary only approach by using the three following procedures. Firstly an integration by parts is performed in the functional, transforming the volume integral into a boundary and a different type of volume integral. Secondly the independent domain variable (potential and displacement in, respectively, potential and elasticity problems) is approximated as a series of products between fundamental solutions and some unknown parameters. With this approximation, the volume integral resulting after the integration by parts becomes equal to zero. The formulation is then written in terms of boundary integrals only, except for the volume integral which appears when there are internal sources (potential case) or body forces (elasticity case). Finally the boundary is subdivided into elements over which the two independent boundary variables are approximated through interpolation functions as it is usually done in either the finite or the classical boundary element methods.

As a consequence of these features the hybrid-displacement BEM presents some interesting characteristics, which will be described in what follows:

- The final system of equations involves matrices whose coefficients are evaluated by performing only boundary integrals (except when internal sources or body forces are present).

- The system of equations involves only one type of boundary variable (potential or displacement) as unknowns, by expressing all the other independent variables in terms of potentials or displacements. Thus a stiffness type of formulation is obtained. The corresponding system matrix is a boundary hybrid stiffness matrix that is obtained as the product of matrices, which are in turn evaluated by boundary integrations only. This boundary stiffness matrix is symmetric and this property can be exploited when necessary. As an example, it can be used in the coupling with a finite element displacement model, which is a simple task due to the stiffness type of the formulation.

- Results for internal points can be obtained easily because no integration is required for their evaluation. This fact might be useful when extending the formulation to plasticity problems, where the computation of internal values is important and has to be done frequently.

- Either concentrated or distributed surface loads can be introduced directly as boundary conditions, whereas in the conventional BEM distributed loads are used.

- The consideration of symmetry is possible without the need for discretization on the axis of symmetry, by applying the so-called implicit symmetry technique, also used in the conventional boundary elements [64].

- The volume integrals which are due to the presence of internal sources (potential) or body forces (elasticity) also occur in the conventional BEM. Therefore the procedures already developed to take these integrals to the boundary can also be used here.

The numerical performance of the new approach is excellent. In all the cases tested, the numerical solutions converge towards the exact results. The accuracy has also been shown to be very good, especially when dealing with the constant element. In potential problems with Dirichlet boundary conditions, the hybrid-displacement BEM has proved to be very accurate, for both constant and quadratic elements. The hybrid solutions for these cases are much more accurate than the conventional BEM solutions for the same problem, when using the same type of element.

The hybrid constant element has shown a very good performance even in the solution of problems where the conventional constant element has proved not to be appropriate. Such is the case of the bending of a plate, shown in the previous chapter, for which the hybrid constant element produced accurate results when a reasonable number of elements were used.

The major drawback of the proposed hybrid formulation is that it is very time-consuming. The high demand of computer time is due to the need to evaluate integrals all along the boundary in the computation of each coefficient of the matrix $\mathbf{F}$. The inversion of the matrix $\mathbf{G}$ necessary to obtain the final stiffness matrix also contributes to increase the computer time. The inversion of matrix $\mathbf{G}$ might be avoided if the main diagonal of the matrix $\mathbf{F}$ could be obtained without the use of the rigid body motion technique. This is a subject for further research.

Another drawback in the formulation is the need for a fine mesh (in comparison to conventional BEM meshes) when using quadratic elements. The hybrid quadratic element has shown to be stiffer than the conventional BEM quadratic element in some problems, in both the potential and the elasticity cases. According to Tong [1, 91], hybrid-displacement finite element models become more flexible if polynomials of lower order are used for the tractions than for the displacements. Based on this statement, in the case of this hybrid BEM formulation, the results for the quadratic element might be improved by using a lower order polynomial for the tractions.

Finally, because this is a different formulation, it can be used to check solutions obtained by the other boundary element approaches. It can also be used together with the other boundary element formulations in the establishment of bench marks for boundary element methods.

Bibliography

[1] Tong, P., New Displacement Hybrid Finite Element Models for Solid Continua, *Int. J. Num. Meth. Eng.*, vol. 2, 1970, pp. 73-83.

[2] Rayleigh, J. W., *The Theory of Sound*, Dover, New York, 1945.

[3] Ritz W., Uber eine neue Methode zur Lösung gewisser Variationsprobleme der mathematischen Physik, *J. reine angew. Math.* , vol. 135, 1908, pp. 1-61.

[4] Ritz W., Theorie der Transversalschwingungen einer quadratischen Platte mit freien Rändern, *Annalen der Physik*, vol. 38, 1909, pp. 737-786.

[5] Courant, Variational Methods for the Solution of Problems of Equilibrium and Vibrations, *Bull. Am. Math. Soc.*, vol. 49, 1943.

[6] Hellinger, E., Die allgemeinen Ansätze der Mechanik der Kontinua, *Enz. math. Wis.*, vol. 4, 1914, pp. 602-694.

[7] Reissner, E., On a Variational Theorem in Elasticity, *J. Math. Phys.*, vol. 29, 1950, pp. 90-95.

[8] Hu, H. C. On some Variational Principles in the Theory of Elasticity and the Theory of Plasticity, *Scientia Sinica*, vol. 4, 1955, pp. 33-54.

[9] Washizu, K., On the Variational Principles of Elasticity and Plasticity, *Tech. Rept. 25-18, Cont. N5ori-07833*, MIT, March 1955.

[10] Nemat-Nasser, S., General Variational Principles in Nonlinear and Linear Elasticity with Applications, *Mechanics Today*, vol. 1, 1972, pp. 214-261.

[11] Prager, W., Variational Principles of Linear Elastostatics for Discontinuous Displacements, Strains and Stresses , *Recent Progress in Applied Mechanics*, The Folke-Odqvist Volume (B. Broberg, J. Hult and F. Niordson Eds.), Almqvist and Wiksell, Stockholm, 1967, pp. 463-474.

[12] Prager, W., Variational Principles for Elastic Plates with Relaxed Continuity Requirements, *International Journal of Solids and Structures*, vol. 4, no. 9, 1968 pp. 837-844.

[13] Washizu, K., *Variational Methods in Elasticity and Plasticity*, Pergamon Press, 1982.

[14] Argyris, J.H. and Kelsey, S., *Energy Theorems and Structural Analysis*, Butterworth, London, 1960.

[15] Turner, M.J., Clough, R. W., Martin, H.C. and Topp, L. J.: Stiffness and Deflection Analysis of Complex Structures, *J. Aeron. Sci.*, vol. 23, no. 9, 1956, pp. 805-824.

[16] Melosh, R. J., Basis for Derivation of Matrices for the Direct Stiffness Method, *J. Am. Inst. Aeron. Astron.*, vol. 1. no. 7, 1963, pp. 1631-1637.

[17] Jones, R. E., Generalization of the Direct Stiffness Method of Structural Analysis, *J. Am. Inst. Aeron. Astron.*, vol. 2, 1964, pp. 821-826.

[18] Pian, T. H. H., Derivation of Element Stiffness Matrices, *J. Am. Inst. Aeron. Astron.*, vol. 2, 1964, pp. 576-577.

[19] Pian, T. H. H., Reflections and Remarks on Hybrid and Mixed Finite Element Methods, *Hybrid and Mixed Finite Element Methods*, Chapter 29 (S. N. Atluri, R. H. Gallagher and O. C. Zienkiewicz Eds.), John Wiley and Sons Ltd, 1983.

[20] Martin, H. C and Carey, G. F., *Introduction to Finite Element Analysis Theory and Application*, McGraw-Hill, New York, 1973.

[21] Pian, T. H. H., Basis of Finite Element for Solid Continua, *Int. J. Num. Meth. Eng.*, vol. 1, 1969, pp. 3-28.

[22] Pian, T. H. H. and Tong, P., Mixed and Hybrid Finite Element Methods, *Finite Element Handbook*, Chapter 5, part 2 (H. Kardestuncer and D. H. Norrie Eds.), McGraw-Hill, New York, 1987.

[23] Noor, A. K., Multifield (Mixed and Hybrid) Finite Element Models, *State-of-the-Art Surveys on Finite Element Technology*, Chapter 5 (A. K. Noor and W. D. Pilkey Eds.), ASME, New York, 1983.

[24] Fredholm, I. Sur une Classe d'Equation Fonctionelles, *Acta Math.*, vol.27, 1903, pp. 365-390.

[25] Kellogg, O. D., *Foundations of Potential Theory*, Dover, New York, 1953.

[26] Muskhelishvili, N. I., *Some Basic Problems of the Mathematical Theory of Elasticity*, P. Noordhoff Ltd., Groningen, 1953.

189

[27] Kupradze, O. D., *Potential Methods in the Theory of Elasticity*, Daniel Davey and Co., New York, 1965.

[28] Hess, J. L. and Smith, A. M. O., Calculation of Potential Flow about Arbitrary Bodies, *Progress in Aeronautical Sciences*, vol. 8 (D. Kuchemann Ed.), Pergamon Press, 1967.

[29] Jaswon, M. A. and Ponter, A. R. S., An Integral Equation Solution of the Torsion Problem, *Proc. Roy. Soc. A*, vol. 273, 1963, pp. 237-246.

[30] Jaswon, M. A., Integral Equation Methods in Potential Theory, I, *Proc. Roy. Soc. A*, vol. 275, 1963, pp. 23-32.

[31] Symm, G. T., Integral Equation Methods in Potential Theory, II, *Proc. Roy. Soc. A*, vol. 275, 1963, pp. 33-46.

[32] Massonnet, C. E., Numerical Use of Integral Procedures in Stress Analysis, *Stress Analysis*, (O. C. Zienkiewicz and G. S. Holister Eds.), Wiley, 1966.

[33] Rizzo, F. J. An Integral Equation Approach to Boundary Value Problems of Classical Elastostatics, *Quart. App. Math.*, vol. 25, 1967, pp. 83-95.

[34] Cruse, T. A., Numerical Solutions in Three Dimensional Elastostatics, *Int. J. Solids Structures*, vol. 5, 1969, pp. 1259-1274.

[35] Rizzo, F. J, and Shippy, D. J., A Method for Stress Determination in Plane Anisotropic Elastic Bodies, *J. Composite Materials*, vol.4, 1970, pp. 36-60.

[36] Cruse, T. A. and Rizzo, F. J., A Direct Formulation and Numerical Solution of the General Transient Elastodynamic Problem, I, *J. Math. Analysis and Applications*, vol. 22, 1968, pp. 244-259.

[37] Cruse, T. A., A Direct Formulation and Numerical Solution of the General Transient Elastodynamic Problem, II, *J. Math. Analysis and Applications*, vol. 22, 1968, pp.341-355.

[38] Cruse, T. A., Application of the Boundary Integral Equation Method to Three Dimensional Stress Analysis, *Computer and Structures*, vol 3., 1973, pp. 509-527.

[39] Brebbia, C. A. and Dominguez, J., Boundary Element Method for Potential Problems, *Appl. Math. Modelling*, vol. 1, 1977, pp. 372-378.

[40] Brebbia, C. A, *The Boundary Element Method for Engineers*, Pentech Press, London, Halstead Press, New York, 1978.

[41] Felippa, C. A., Parametrized Multifield Variational Principles in Elasticity: I. Mixed Funcionals, *Communications in Applied Numerical Methods*, vol. 5, 1989, pp. 79-88.

[42] Felippa, C. A., Parametrized Multifield Variational Principles in Elasticity: II. Hybrid Funcionals and the Free Formulation *Communications in Applied Numerical Methods*, vol. 5, 1989, pp. 89-98.

[43] McDonald, B. H., Friedman, M. and Wexler, A., Variational Solution of Integral Equations, *IEEE Trans. Microwave Theory Tech.*, vol. MTT-22, 1974, pp. 237-248.

[44] Jeng, G. and Wexler, A., Isoparametric, Finite Element, Variational Solution of Integral Equations for Three-Dimensional Fields, *Int. J. Num. Meth. Eng.*, vol. 11, 1977, pp. 1455-1471.

[45] Jeng, G. and Wexler, A., Self Adjoint Variational Formulation of Problems Having Non-Self-Adjoint Operators, *IEEE Trans. Microwave Theory Tech.*, vol. MTT-26, 1978, pp. 91-94.

[46] Hsiao, G. C., Kopp, P. and Wendland, W. L., The Synthesis of the Collocation and the Galerkin Method Applied to Some Integral Equations of the First Kind, *New Developments in Boundary Element Methods*, CMC Publication, Southampton, England, 1980, pp. 122-136.

[47] Wendland, W. L., On the Asymptotic Convergence of Boundary Integral Methods, *Boundary Element Methods, Proc. 3rd Int. Seminar on BEM in Engineering* (C. A. Brebbia Ed.), Southampton, England, 1981.

[48] Wendland, W. L., Asymptotic Accuracy and Convergence, *Progress in Boundary Element Methods* (C. A. Brebbia Ed.), Pentech Press, London, vol. 1, 1981, pp. 289-313.

[49] Nedelec, J. C., Integral Equations with Non-Integrable Kernels, *Integral Equations and Operator Theory*, vol. 5, 1982, pp. 562-572.

[50] Arnold, D. N. and Wendland, W. L., Collocation versus Galerkin Procedures for Boundary Integral Methods, *Boundary Element Methods in Engineering, Proc. 4rd Int. Seminar on BEM in Engineering* (C. A. Brebbia Ed.), Southampton, England, 1982.

[51] Polizzotto, C., A BEM Approach to the Bounding Techniques, *Boundary Elements VII* (C. A. Brebbia Ed.), Springer-Verlag, Berlin, 1985, pp. 13/103-114.

[52] Maier, G. and Polizzotto, C., A Galerkin Approach to Elastoplastic Analysis by Boundary Elements, *Comput. Meth. Appl. Mech. Eng.*, vol. 60, 1987, pp. 175-194.

[53] Onishi, K., Heat Conduction Analysis Using Single Layer Heat Potential, *Boundary Element Methods in Applied Mechanics* (M. Tanaka and T. A. Cruse Eds.), Pergamon Press, 1988.

[54] Dumont, N. A., The Variational Formulation of the Boundary Element Method, *Boundary Element Techniques: Applications in Fluid Flow and Computational Aspects* (C. A. Brebbia and W. S. Venturini Eds.), Computational Mechanics Publications, Southampton, 1987.

[55] Dumont, N. A., The Hybrid Boundary Element Method, *Boundary Elements IX* (C. A. Brebbia, W. L. Wendland and G. Kuhn Eds.), vol. 1, Springer-Verlag, Berlin, 1987, pp. 117-130.

[56] DeFigueiredo, T.G.B. and Brebbia, C.A., A Hybrid Displacement Variational Formulation of BEM, *Boundary Elements X* (C.A. Brebbia Ed.), vol. 1, Computational Mechanics Publications, Southampton, Springer-Verlag, Berlin, 1988, pp. 33-42.

[57] DeFigueiredo, T.G.B. and Brebbia, C.A., A New Hybrid Displacement Variational Formulation of BEM for Elastostatics, *Advances in Boundary Elements* (C.A. Brebbia and J.J. Connor Eds.), vol. 1, Computational Mechanics Publications, Southampton, Springer-Verlag, Berlin, 1989, pp. 47-57.

[58] Polizzotto, C., A Consistent Formulation of the BEM within Elasto-plasticity, *Advanced Boundary Element Methods* (T. A. Cruse Ed.), Springer-Verlag, Berlin, 1987, pp. 315-324.

[59] Polizzotto, C. and Zito, M., A Variational Approach to Boundary Element Methods, *Boundary Element Methods in Applied Mechanics* (M. Tanaka and T. A. Cruse Eds.), Pergamon Press, Oxford, 1988.

[60] Felippa, C. A., A Parametric Representation of Symmetric Boundary Finite Elements, *Topics in Engineering vol. 7* (M. Tanaka and R. Shaw Eds.), Computational Mechanics Publications, 1990.

[61] Jaswon, M. A. and Symm, G. T., *Integral Equation Methods in Potential Theory and Elastostatics*, Academic Press, London, 1977.

[62] Rao, S. S. *The Finite Element Method in Engineering*, Pergamon Press, Oxford, 1982.

[63] Huebner, K. H., *The Finite Element Method for Engineers* , John Wiley and Sons, New York, 1975.

[64] Brebbia, C. A., Telles, J. C. F. and Wrobel, L. C., *Boundary Element Techniques*, Springer Verlag, Berlin, 1984.

[65] Brebbia,C. A. and Dominguez, J., *Boundary Elements An Introductory Course*, Computational Mechanics Publications, Southampton and McGraw-Hill, New York, 1989.

[66] Kaplan, W., *Advanced Mathematics for Engineers*, Addison-Wesley, Reading - Massachussets, 1981.

[67] Stroud, A. H. and Secrest, D., *Gaussian Quadrature Formulas*, Prentice-Hall, New York, 1966.

[68] Telles, J. C. F., A Self-Adaptive Coordinate Transformation for Efficient Numerical Evaluation of General Boundary Element Integrals, *Int. J. Num. Meth. Eng.*, vol. 24, 1987, pp. 969-973.

[69] Telles, J. C. F., The Boundary Element Method Applied to Inelastic Problems (C. A. Brebbia and S. A. Orzag Eds.), Springer-Verlag, Berlin, 1983.

[70] Love, A. E. H., *A Treatise on the Mathematical Theory of Elasticity*, Dover Publications, New York, 1944.

[71] Melan, E., Der Spannungszustand der durch eine Einzerkraft im Innern beanspruchten Halbscheibe, *Z. Angew. Math. Mech.*, vol. 12, 1932, pp. 343-346.

[72] Midlin, R. D., Force at a Point in the Interior of a Semi-infinite Solid, *Physics*, vol. 7, 1936, pp. 195-202.

[73] Georgiou, P., *The Coupling of the Direct Boundary Element Method with the Finite Element Displacement Technique in Elastostatics*, PhD Thesis, University of Southampton, England, 1981.

[74] Tulberg, O. and Bolteus, L., A Critical Study of Different Boundary Element Stiffness Matrices, *Proc. 4th Int. Seminar on BEM in Engineering* (C. A. Brebbia Ed.), Southampton, England, 1982.

[75] Kutt, H. R., The Numerical Evaluation of Principal Value Integrals by Finite Part Integration, *Numer. Math.*, vol. 24, 1975, pp. 205-210.

[76] Kutt, H. R., Quadrature Formulae for Finite Part Integrals, *Report WISK 178*, The National Research Institute for Mathematical Sciences, Pretoria, 1975.

[77] Rizzo, F. J, and Shippy, D. J., An Advanced Boundary Integral Equation Method for Three Dimensional Thermoelasticity, *Int. J. Num. Meth. Eng.*, vol. 11, 1977, pp. 1753-1768.

[78] Stippes, M. and Rizzo, F. J., A Note on the Body Forces Integral of Classical Elastostatics, *Z. Angew. Math. Phys.*, vol. 28, 1977, pp. 339-341.

[79] Danson, D. J., A Boundary Element Formulation of problems in Linear Isotropic Elasticity with Body Forces, *Boundary Element Methods* (C. A. Brebbia Ed.), Springer-Verlag, Berlin, 1981, pp. 105-122.

[80] Telles, J. C. F. and Brebbia, C. A., The Boundary Element Method in Plasticity, *New Developments in Boundary Element Methods* (C. A. Brebbia Ed.), CML Publications, Southampton, 1980, pp. 295-317; *Appl. Math. Modelling*, vol. 5, 1981, pp.275-281.

[81] Malvern, L. E., *Introduction to the Mechanics of a Continuous Medium*, Prentice-Hall, Englewood Cliffs, N. J., 1969.

[82] Fung, Y. C., *Foundations of Solid Mechanics*, Prentice-Hall, Englewood Cliffs, N. J., 1965.

[83] Cruse, T. A., Mathematical Foundations of the Boundary Integral Equation Method in Solid Mechanics, *Report No. AFOSR-TR-77-1002*, Pratt and Witney Aircraft Group, 1977.

[84] Cruse, T. A., Boundary Integral Equation Method for Three- dimensional Elastic Fracture Mechanics, *Report No. AFOSR-TR-75-0813*, Pratt and Witney Aircraft Group, 1975.

[85] Paula, F. A., *Obtenção de Matriz de Rigidez Utilizando o Método dos Elementos de Contorno*, MSc. Thesis, COPPE/UFRJ, Rio de Janeiro, Brazil, 1986 (in Portuguese).

[86] Fairweather, G. et al., On the Numerical Solution of Two-Dimensional Potential Problems: an Improved Boundary Integral Equation Method, *J. of Computational Physics*, vol. 31, 1979, pp. 96-112.

[87] Greenspan, D., *Introductory Numerical Analysis of Elliptic Boundary Value Problems*, Harper & Row, New York, 1965.

[88] Papamichael, N. and Symm, G. T., Numerical Techniques for Two-Dimensional Laplacian Problems, *Comput. Meth. Appl. Mech. Eng.*, vol. 6, 1975, pp. 175-194.

[89] Timoshenko, S.P. and Goodier, J.N., *Theory of Elasticity*, McGraw-Hill, 1982.

[90] Roark, R.J. and Young, W.C., *Formulas for Stress and Strain*, McGraw-Hill, London, 1975.

[91] Tong, P., Basis of Finite Element Methods for Solid Continua, *Int. J. Num. Meth. Eng.*, vol. 1, 1969, pp. 3-28.

[92] Mangler, K. W., Improper Integrals in Theoretical Aerodynamics, *Report No: Aero.2424*, Royal Aircraft Establishment, 1951.

NOTATION

∇^2	Laplacian operator
δ_{ij}	Kronecker delta
$\Delta(\xi, x)$	Dirac delta function
Ω	domain
Γ	boundary
$\mathbf{n}$	outward normal unit vector
x	field point
ξ	source point
u	potential
q	normal flux
u^*	fundamental solution for potential
q^*	normal flux corresponding to the potential fundamental solution
u_i	displacement vector component
p_i	traction vector component
u_{ij}^*	component of the tensor of fundamental solution for displacement
p_{ij}^*	component of the tensor of tractions corresponding to the fundamental solution
σ_{ij}	stress tensor component
ϵ_{ij}	stress tensor component

K	stiffness matrix
F, G, L	matrices used to evaluate the stiffness matrix
~	indicates boundary variable
–	indicates prescribed variable

Lecture Notes in Engineering

Edited by C. A. Brebbia and S. A. Orszag

Vol. 40: R. Borghi, S. N. B. Murhty (Eds.)
Turbulent Reactive Flows
VIII, 950 pages. 1989

Vol. 41: W. J. Lick
Difference Equations
from Differential Equations
X, 282 pages. 1989

Vol. 42: H. A. Eschenauer, G. Thierauf (Eds.)
Discretization Methods
and Structural Optimization –
Procedures and Applications
Proceedings of a GAMM-Seminar
October 5-7, 1988, Siegen, FRG
XV, 360 pages. 1989

Vol. 43: C. C. Chao, S. A. Orszag, W. Shyy (Eds.)
Recent Advances in Computational
Fluid Dynamics
Proceedings of the US/ROC (Taiwan) Joint
Workshop in Recent Advances in
Computational Fluid Dynamics
V, 529 pages. 1989

Vol. 44: R. S. Edgar
Field Analysis and
Potential Theory
XII, 696 pages. 1989

Vol. 45: M. Gad-el-Hak (Ed.)
Advances in Fluid Mechanics
Measurements
VII, 606 pages. 1989

Vol. 46: M. Gad-el-Hak (Ed.)
Frontiers in Experimental
Fluid Mechanics
VI, 532 pages. 1989

Vol. 47: H. W. Bergmann (Ed.)
Optimization: Methods and Applications,
Possibilities and Limitations
Proceedings of an International Seminar
Organized by Deutsche Forschungsanstalt für
Luft- und Raumfahrt (DLR), Bonn, June 1989
IV, 155 pages. 1989

Vol. 48: P. Thoft-Christensen (Ed.)
Reliability and Optimization
of Structural Systems '88
Proceedings of the 2nd IFIP WG 7.5 Conference
London, UK, September 26-28, 1988
VII, 434 pages. 1989

Vol. 49: J. P. Boyd
Chebyshev & Fourier Spectral Methods
XVI, 798 pages. 1989

Vol. 50: L. Chibani
Optimum Design of Structures
VIII, 154 pages. 1989

Vol. 51: G. Karami
A Boundary Element Method for
Two-Dimensional Contact Problems
VII, 243 pages. 1989

Vol. 52: Y. S. Jiang
Slope Analysis Using
Boundary Elements
IV, 176 pages. 1989

Vol. 53: A. S. Jovanovic,
K. F. Kussmaul, A. C. Lucia,
P. P. Bonissone (Eds.)
Expert Systems in Structural
Safety Assessment
X, 493 pages. 1989

Vol. 54: T. J. Mueller (Ed.)
Low Reynolds Number
Aerodynamics
V, 446 pages. 1989

Vol. 55: K. Kitagawa
Boundary Element Analysis
of Viscous Flow
VII, 136 pages. 1990

Vol. 56: A. A. Aldama
Filtering Techniques for
Turbulent Flow Simulation
VIII, 397 pages. 1990

Vol. 57: M. G. Donley, P. D. Spanos
Dynamic Analysis of Non-Linear
Structures by the Method of
Statistical Quadratization
VII, 186 pages. 1990

Vol. 58: S. Naomis, P. C. M. Lau
Computational Tensor Analysis
of Shell Structures
XII, 304 pages. 1990